LESSER
BEASTS

LESSER

BEASTS

A Snout-to-Tail History of the Humble Pig

Mark Essig

BASIC BOOKS

A Member of the Perseus Books Group
New York

Published by Basic Books,
A Member of the Perseus Books Group

Books published by Basic Books are available at special discounts for bulk
purchases in the United States by corporations, institutions, and other
organizations. For more information, please contact the Special Markets
Department at the Perseus Books Group, 2300 Chestnut Street, Suite 200,
Philadelphia, PA 19103, or call (800) 810-4145, ext. 5000, or e-mail
special.markets@perseusbooks.com.

Designed by Pauline Brown

Library of Congress Cataloging-in-Publication Data
    Essig, Mark, 1969–
        Lesser beasts : a snout-to-tail history of the humble pig / Mark Essig.
        pages cm
        Includes index.
        ISBN 978-0-465-05274-5 (hardcover : alk. paper)—
    ISBN 978-0-465-04068-1 (e-book) 1. Swine—History. 2. Pork—
    History. I. Title.
    SF395.E64 2015
    636.4—dc23
                                                2014049256

10 9 8 7 6 5 4 3 2 1

For Melissa, Jack, and Lydia

"Cattle country" calls up instant visions of distant mountains and wind-swept plains leading off to nowhere and cattle grazing on slopes and tattooed men in wide-brim hats gathered around a fire with their horses standing stalwartly in the background. . . . But who, among the teeming city masses, knows about "hog country"? Who knows where lies that land?

—William Hedgepeth, *The Hog Book*

"Humble?" said Charlotte. "'Humble' has two meanings. It means 'not proud' and it means 'near the ground.' That's Wilbur all over. He's not proud and he's near the ground."

—E. B. White, *Charlotte's Web*

CONTENTS

THE MAGICAL ANIMAL

On a trip through the North Carolina mountains in 1878, Virginia newspaper editor James Cowardin found himself surrounded by thousands of pigs. "Hogs were before us and behind us, and both to the right and to the left of us," Cowardin wrote. "There was whipping and shouting and twisting and turning" as the swineherds yelled, "Suey!" "Suey!" "Get out!" "Suey hogs!" "D—d devil take the swine!" Cowardin too cursed the pigs at first, but once he settled into the rhythm of the road, he began to daydream about following his "grunting friends" to their destination and enjoying a pig slaughter feast: "What luxury in spare ribs, backbone, and sausage we would have," he fantasized, "not to mention pigs' tails broiled on hot rocks!"

The flesh of Cowardin's traveling companions, though, was destined for other stomachs. He had stumbled upon a seasonal movement of livestock that had been happening each winter for half a century. The swine had been fattened in eastern Tennessee, a fertile farming region with many pigs and few people. A couple of hundred miles away lay the plantations of the South, which didn't raise much food. Planters preferred to grow cotton, sell it for cash, and buy pork to feed their slaves (or, after the Civil War, their sharecroppers and tenant farmers). The hog supply was in Tennessee, the demand in South Carolina and Georgia, and in between lay the Blue Ridge mountains. No rivers or railroads connected the two, so there was only one way to move the hogs: on foot.

Hog droving, as the practice was known, formed an essential link in the global economy. In peak years as many as 150,000 hogs made the journey on this single turnpike, and many other mountain routes also carried pigs from upland farms to the Deep South. The pork fed the slaves, who raised the cotton, which supplied textile mills in New England and Great Britain, which made the fabric that clothed the world. And it all depended on a few men herding hogs through a narrow river valley cutting through the mountains of North Carolina.

I first learned of hog drives in 2007, not long after my family and I settled in those mountains. A historical marker revealed that "livestock drovers" once traversed a road near our home in Asheville, North Carolina. I had thought cattle drives happened on the Great Plains, not in the mountains, so I headed to the library, where I read books on local history, scanned microfilm of nineteenth-century newspapers, and searched Google Books for old runs of defunct farming magazines. And I discovered the strange truth: most of the animals herded through Asheville had been not cows or even sheep but pigs.

The story of these pigs, I learned, was even etched into the landscape: a local farmer showed me a spot on his land where an old drovers' road is still visible, a deep trench cut into the clay soil by decades of wagon wheels and sharp little hooves. Think of it: pig drives! Like cattle drives, only stranger! Who knew a pig could walk that far or would travel in the desired direction? Apparently not many people: I read a 2006 article by a prominent archaeologist, a specialist in livestock, who baldly insisted that pigs "cannot be driven." The historical record suggests that pigs can indeed be driven. In fact, if you gave them a few lessons and a specially designed steering wheel, I wouldn't be surprised if pigs could drive.

At about this time I started teaching journalism at Warren Wilson College, a liberal arts school that also operates a farm. The animals live on pasture, and nutrients cycle from the soil into crops, from crops into the mouths of animals, and from animal manure back into the soil. I observed this cycle firsthand one day each week when I volunteered on the pig crew. Working alongside students, I scraped manure, topped up feeders, clipped the milk teeth of newborn piglets, and castrated the young males. I spent a lot of time just watching: a dominant sow chasing off her weaker sisters to get first dibs at the trough; enormous boars, rendered bowlegged by their cantaloupe-sized testicles, hoisting themselves atop sows in heat; young pigs scattering across the pasture as I approached, then returning to sniff and prod at my boots with their snouts. A boar known as Gucci—the students made the most of their naming duties— would prop his front legs on the wall of his pen and gaze around the farmyard contentedly, a lord surveying his estate. The pigs were by turns curious, surly, skittish, and playful. They were the most fascinating creatures in the barnyard, brainy and fully alive.

I headed back to the library and began following the trail of the pig around the world and back into prehistory. I met

all kinds of swine: wild boar that lurked around Neolithic
villages to scavenge garbage and gradually domesticated
themselves; the outcast pigs of ancient Israelites and their
neighbors, rejected as unclean; the beloved fat white swine of
the Roman Empire, sacrificed to the gods and roasted whole
for banquets; the rangy forest hogs of medieval Europe and
colonial America, thriving under conditions of utter neglect
and helping pioneers tame new land; the low-bellied pigs of
the Chinese, kept in tiny sties, growing fat on rice bran and
other farm wastes; the urban pigs of England and America,
living in backyard sties or roaming city streets, providing
the poor with their only source of meat; the hybrid Chinese-
European swine of the nineteenth-century Corn Belt, turning
corn into meat on a tight ratio and providing protein for an
urbanizing nation; and, finally, the pigs of modern agribusiness,
raised in windowless metal sheds, dining on a precisely cal-
ibrated blend of corn, soy, and antibiotics, producing cheap
meat to feed the world.

The 10,000-year history of the domestic pig is a tale of both
love and loathing. A prodigious producer of meat—chubby bul-
wark against human malnutrition, centerpiece of medieval feasting
and southern barbecue, precious mother-source of bacon—the pig
has just as often met with contempt. For thousands of years,
many people have either refused pork entirely or approached it
with extreme caution.

The problem of the pig seems especially relevant today. At a
time when choosing food is more complicated than ever—when
buying a pork chop raises thorny questions about the environment, public health, workers' rights, and animal welfare—
it makes sense to take a look back at what has been, for several
thousand years, the most controversial of foods. Why do pigs
provoke feelings of disgust? Why have so many people rejected

pork? The answers to those questions lie deep in the past, tangled up in the biology of people and pigs, in shifting environmental and economic conditions, and in the ways people find meaning in the foods they eat.

Pigs "were generally recognized as being the cleverest of animals," George Orwell writes in *Animal Farm*, where the pigs take charge of the barnyard and declare themselves "more equal than" their fellow beasts. Science justifies that arrogance. Studies show that pigs can figure out how mirrors work and use them to scan the landscape for a meal. A pig that knows where food is cached will delay its gratification until no other pigs are present and then enjoy the meal by itself. It can learn to perform tasks—open a cage, turn a heater on and off, play video games—more quickly than nearly any other animal. Animal scientist Temple Grandin reports that in barns that use electronic collars to dispense individual portions of food, sows who find a stray collar on the ground will carry it to the food dispenser to steal a second helping.

The ancients recognized pig intelligence. Pliny the Elder claimed that pigs aboard a listing ship would scramble to the higher side to balance the cargo. Sows have been trained to hunt truffles since Babylonian times, though they have a habit of rooting up the prize for their own enjoyment. In early nineteenth-century England, a black sow bearing the unfortunate name Slut—the word also meant "filthy"—worked as a pointer alongside her owner's hunting dogs. At about the same time, London theaters staged performances by trained pigs who told the time, spelled words, and solved math problems. "Pigs are a race unjustly calumniated," Samuel Johnson observed. "We do not allow time for his education; we kill him at a year old."

Toby was one of many "learned pigs" who spelled words and solved math problems onstage in England and America in the eighteenth and nineteenth centuries. Such performances functioned as burlesque—the lowly swine displaying higher mental powers—but also betrayed an anxiety that the beasts we eat might be nearly as smart as we are.

The pig's intelligence, however, generally failed to alter its fate. In one twentieth-century vaudeville act, the pigs at a certain point would refuse to do tricks. The trainer, dressed as a butcher, began sharpening a large knife, whereupon the pigs did a double-take and sprang back into action. The joke cut close to home: nearly all such performers made their final appearance on the dinner table. The first pig to play Arnold Ziffel, star of the television series *Green Acres*, ended up as pork chops in his trainer's freezer.

Cleverness has never been the pig's primary value to human beings. Nearly every society has placed a high value on meat. In one convenient package, it provides high-quality protein, fat, vitamins, and trace minerals, all necessary for survival.

Our bodies crave meat, and our minds scheme to acquire it. Although humans can satisfy all their nutritional needs by eating plants, most of us prefer not to. Instead, we feed plants to animals and then eat the animals, and we do not seem to mind that this process is costly and complicated. People have fought wars, conquered lands, destroyed landscapes, and exchanged great wealth to satisfy their deep hunger for meat. When it is scarce—and in large societies meat has been scarce until recent times—only the wealthiest eat it. When poor people begin to earn a bit more money, they spend it on meat. "Those who could, gorged themselves," one historian has written of early modern Europe. "Those who couldn't, aimed to."

More often than not, the most readily available meat was pork. That was due partly to the pig's versatile diet. Whereas cows and sheep must live on pasture, eating grass, pigs, like people, are omnivores. They will eat corn in the field, garbage on city streets, kitchen slop in backyard sties, whey in dairy barns, acorns in forests, and mollusks on tropical beaches.

Self-sufficiency added to the pig's appeal. Give pigs plenty of food and they'll loll about the sty and grow fat. Take the food away and they'll slip into the woods and fend for themselves. When explorers in the sixteenth century encountered uninhabited islands, they would drop a boar and a few sows on shore—sort of like tossing a handful of seeds into a jungle and expecting a vegetable garden to grow. Except that it worked. Left alone, the island pigs survived and multiplied, providing a bountiful food supply for the next passing ship. Swine served the same function closer to home. Well into the twentieth century, many American farmers turned their pigs loose in the woods, where the animals fed themselves until they were rounded up for slaughter. Those that escaped the roundup began to live and breed in the woods like wild animals, creating a thriving population of feral pigs. The United States is now home to an estimated 5 million feral swine that ravage crops, undermine levees, devour rare salamanders, and root up the turf on golf courses. Efforts to control them—including shooting them from helicopters—have proved futile because pigs breed so quickly.

The same quality that makes feral pigs a problem—prodigious fecundity—has delighted farmers. Cows, goats, and sheep provide milk, a bountiful and consistent source of protein. Oxen pull plows and carts, and sheep are shorn for wool. Pigs do not pull plows; they give no milk and grow no wool. Pigs produce only one thing: more pigs. Many, many more pigs. Cows gestate for nine months and produce one calf; sheep and goats require five months and give birth to one or two offspring. A sow, on the other hand, gestates for less than four months and produces eight or twelve or even more piglets, all of which grow to slaughter weight far more quickly than a calf or a lamb. Born at 3 pounds, today's piglet balloons to 280 pounds by six months of age, at which point it is also ready to breed. In 1699 a French scholar estimated that in one decade—even making allowances for illness and predation by wolves—a single sow could become grandmother to 6 million pigs. The calculation was perhaps optimistic, but it carried its point.

All of those pigs were good for only one thing: meat. The two-line poem "Bacon & Eggs," attributed to Howard Nemerov, captures this uncomfortable fact:

> The chicken contributes,
> But the pig gives his all.

The pig's certain doom has launched the plot of many a children's tale: a cow earns its keep giving milk, but a pig saves itself only by developing an oddball talent such as herding sheep or inspiring a spider to write words in her web.

In the unsentimental realm of real-world farms, pigs eat, grow fat, and get killed. This process happens quickly and efficiently, and the result is meat that tastes delicious either fresh or preserved with salt and smoke. Cured beef or mutton often tastes like shoe leather, but pork—as bacon and ham lovers know—only gets better. In the time before artificial refrigeration, which has existed for less than 1 percent of recorded history, it provided a year-round source of protein.

Pork also offers variety. In one episode of *The Simpsons*, Lisa becomes a vegetarian. Homer asks her, "Are you saying you're never going to eat any animal again? What about bacon?"

"No."
"Ham?"
"No."
"Pork chops?"
"Dad! Those all come from the same animal!"

"Oh, yeah, right, Lisa," Homer says with heavy sarcasm, waggling his fingers in the air. "A wonderful, magical animal!"

Homer here echoes an opinion first recorded by Pliny the Elder about a century before Christ. "There is no animal that affords a greater variety to the palate," Pliny wrote in his *Natural History*. "All the others have their own peculiar flavor, but the flesh of the hog has nearly fifty different flavors." Given the abundance, variety, and toothsomeness of this meat, it's easy to understand the enthusiasm of an American farm woman who in 1849 greeted the arrival of the year's first litter with this entry in her diary: "Pigs! Pigs! Pork! Pork! Pork!"

With meat the most valued of foods and pigs a prolific producer of meat, one might assume that pigs have been universally embraced. In Asia and the Pacific Island region, that assumption would be largely correct. Pigs stood at the center of cultural life in much of Polynesia and Micronesia, serving as victims in ritual sacrifice and as the key source of protein; women sometimes suckled orphaned piglets alongside their own children. China's traditional agriculture revolved around pigs as producers of manure, and its cooks prized pork above all other meats. The Chinese character for "home" is formed by placing the symbol for "pig" under the symbol for "roof": home is where the pig is.

In the Western world—which, for reasons of brevity, is the focus of this book—the relationship between pigs and people has been fraught. Judaism placed a complete ban on pork in the first century BC, and Islam followed suit more than 1,000 years later. Christians gave their blessing to pork but still found it difficult to shake off Old Testament prejudice, condemning pigs as lazy, filthy, and gluttonous. Englishmen ranked pork as the least desirable of meats. Americans, from colonial days until after World War II, ate far more pork than beef but nonetheless disparaged pork and insulted the pig. The cow—that great, dumb, placid beast with a thousand-yard stare and only a faint glimmer of intelligence—stole all the glory, with steaks and cowboys central to the mythology of America. In a classic work of American agricultural history, cows get a chapter to themselves, while hogs are consigned to a catch-all chapter titled "The Lesser Beasts."

Greater in dietary significance than the cow, the pig certainly has been lesser in prestige. Foods, like the societies that consume them, are arranged into hierarchies. For those linked by blood, religion, class, or nation, sharing a meal forges bonds but also draws boundaries: we use food to stigmatize foreigners, exclude nonbelievers, climb the social ladder, and kick others down a few rungs. No food has played a bigger role than pork in shaping cultural identities, and examining why this is so might help untangle some of our current dilemmas surrounding food.

Historically, the question of which people eat which foods has been a matter of tradition, price, status, and availability: the rich eat what is rare and expensive, and the poor eat whatever they can afford. Today, the calculations have shifted. Consumers with the means ask themselves if pesticides lurk in the folds of their lettuce, if their chips were made from genetically modified corn, and if the workers who picked their tomatoes were paid a fair wage. The meat counter poses a separate set of dilemmas: Does the chicken harbor antibiotic residues or dangerous bacteria? Did the steak come from a cow fed slaughterhouse by-products? Did the ham require undue suffering on the part of a pig?

People, in other words, are pausing to think about the animals that become meat: What did they eat, where did they live, and how did they die? Similar questions have been asked about pigs for thousands of years.

The anxiety about pigs starts with their omnivorous appetite. In addition to acorns and rice hulls, pigs happily devour that which most disgusts us—rotting garbage, feces, carrion, even human corpses. Of all the animals commonly eaten by humans, the pig is the only one that will return the favor. Many texts, from scripture to Shakespeare, have noted the pig's willingness to scavenge human bodies, and such incidents happen even today. In 2012 an Oregon farmer went to feed his sows and never returned; authorities later searched the sty and recovered his dentures and a few scattered body parts. If you are what you eat—an age-old expression found in many languages—then what's eaten by the animals you eat becomes cause for concern.

Pigs became pariahs in Egypt, Mesopotamia, and Palestine as early as 1000 BC, but even there they didn't disappear. Instead, they took up residence among society's human outcasts, living as scavengers among the homes of the poor. For most of history, the vast majority of people lived in danger of starvation, and only a fortunate few could afford to be picky about the food they ate. The elites who wrote dietary laws and set culinary fashions may have turned up their noses at pigs, but people on the margins embraced them as a nearly free source of food. In the tail-chasing realm of social status, this further damaged the reputation of pigs, who became contemptible not only for their own dirty habits but also because they kept company with the poor.

The scavenging ways of pigs created a strange intimacy. Cattle and sheep, throughout history, have generally ranged on the fringes of settlements. Pigs, by contrast, often spent their days quite near people's homes. They became members of the family, consuming the leftovers of meals before becoming dinner themselves. Eating pigs sometimes seemed to border on cannibalism and required emotional distancing. An old joke tells of a visitor to a farm who spots a pig with a wooden leg. The pig, he learns, has saved the farmer from drowning, scared away a bear, and rescued all the other farm animals from a barn fire. "What an amazing animal," the visitor asks. "But how did he lose his leg?" The farmer responds, "Well, a pig like that, you don't eat him all at once."

Or perhaps you don't eat him at all. In E. B. White's *Charlotte's Web*, Charlotte the spider saves Wilbur the pig from slaughter by weaving a few words in silk. One of those words is "humble," which Charlotte says suits Wilbur because it means "not proud" and "near the ground." She doesn't mention a third meaning: "inferior" or "low class." Are pigs humble? The word shares a root with "humus," or soil, where pigs spend a lot of time rooting. And for much of history they have been considered inferior—filthy animals eaten by low-class people. But "not proud" misses the mark. Wilbur notwithstanding, pigs are fractious, independent minded, and difficult to herd. Given the chance, they'll happily turn the tables and make a meal of a person. There's nothing humble about that.

Compared to cows and sheep, pigs are arrogant bastards. "There's always a certain tension about a bunch of pigs walking around," a twentieth-century hog farmer said. "You never know when they're gonna flare up—start bitin' off another one's ear or something. You just don't get the calm, peaceful feeling like when you see a herd of sheep." His statement reveals frustration but also admiration. It's the voice of a father who loves best his worst-behaved child, who knows that docility is close kin to stupidity, who sees in his hogs a bit of himself.

The line between people and pigs can be fluid. Alice, during her adventures in Wonderland, carries a baby that gradually turns into a piglet, so she sets it down and, with some relief, watches it trot into the woods: "'If it had grown up,' she said to herself, 'it would have made a dreadfully ugly child: but it makes rather a handsome pig, I think.'" In *The Odyssey* the enchantress Circe transforms sailors into swine but leaves them "cursed with sense"—Odysseus's men, that is, retain human minds within animal bodies. The story gives voice to the suspicion that when we eat a pig, we eat our close kin. Renaissance physicians claimed that human flesh tasted like pork, and reports from cannibals have supported that claim.

Few religious rules govern the consumption of vegetables; taboos and regulations cluster around meat. Such rules may

have functioned as public health measures—meat is more likely
to harbor parasites and lethal bacteria—but they also acknowledged the significance of taking life. Killing an animal and eating
its flesh traditionally has been considered a sacred act, which is
why—in ancient Greece, Israel, and many other cultures—the
roles of butcher and priest often blended together: holy men
killed animals at the altar according to sacred protocols, offered
burnt offerings to their God or gods, and then distributed the remaining meat to the crowd. Ritual sanctified the spilling of blood.

It also offered distraction from an uncomfortable fact: the
substance we call meat is virtually identical to the flesh on our
own bones. Pork presents this problem in acute form. People
and pigs share roughly similar teeth, skin, and internal anatomy.
Renaissance doctors dissected pigs as models for humans. Modern surgeons transplant pig heart valves into people. Scientists
are developing genetically modified pigs with "humanized"
lungs for transplantation into people. Pigs get ulcers, arthritis,
and diabetes, just like we do. They're also smart. They like to
watch TV and drink beer, and, given the chance, they tend to
grow fat and sedentary.

Confronted by this uncanny beast, humans have reacted
with a blend of attraction and revulsion, hunger and disgust.
"Dogs look up to you, cats look down on you," Winston Churchill once said. "Give me a pig—he just looks you in the eye and
treats you as an equal." We look back at the pig and see quite
a bit of ourselves. And then, more often than not, we eat him.

ONE

KEEP IT SIMPLE

In late February 1922, Henry Fairfield Osborn, head of New York's American Museum of Natural History, received a molar in the mail. A fossil hunter named Harold Cook had unearthed the tooth while digging in the 10-million-year-old Snake Creek fossil beds of western Nebraska. The tooth, Cook told Osborn, "very closely approaches the human type."

The scientist agreed. "It is the last right upper molar tooth of some higher primate," Osborn told Cook. "We may cool down tomorrow, but it looks to me as if the first anthropoid ape of America has been found."

Osborn did not cool down. A month after first examining the tooth, he published a scientific article proclaiming that, millions of years ago, a human-like primate had walked the plains of North America. The *Illustrated London News* ran a fanciful drawing depicting a brawny, slope-shouldered, club-wielding ape-man. A worldwide mania for "Nebraska Man" commenced.

The timing could not have been better for Osborn, who was just then engaged in a public dispute with William Jennings Bryan, the great populist leader. Bryan had launched a campaign against Darwinism that would culminate a few years later with the Scopes Monkey Trial in Tennessee, which tested a law that banned the teaching of evolution. The *New York Times* explained that the tooth provided "further evidence that Mr. Bryan is wrong and Darwin was right." Even better, the fossil had been discovered in Bryan's home state. Osborn did not let the irony pass unnoted, suggesting in his journal article that the new ape-man "should be named *Bryopithecus* after the most distinguished primate which the State of Nebraska has thus far produced."

The joke soured quickly. Further expeditions turned up more teeth that undermined Osborn's claims. A retraction in the journal *Science* acknowledged that the tooth had come not from a man or an ape but from an extinct piglike creature.

The story of Nebraska Man is remembered today mostly by Bryan's intellectual descendants, the creationists, who claim that Darwinists extrapolate wildly from slight evidence. And they seem to have a point: How could a great scientist confuse a pig tooth with a primate tooth?

As it turns out, the mistake was an easy one to make. The Nebraska fossil came from a peccary, close cousin to the pig. Pigs and peccaries have incisors for cutting, canines for tearing, and premolars and molars for chewing and grinding. The full set closely matches those of people, and that is what got Osborn into trouble. He erred by drawing his conclusion on the basis of an old tooth. Young molars have distinctive cusps that reveal the species of origin. Once those cusps wear away, the molars of pigs and people are nearly identical.

When Osborn confused those teeth, he may have rushed to scientific judgment, but he also exposed an important truth: pigs

The 1922 discovery of a fossilized molar prompted speculation that "Nebraska Man" once roamed the Midwest. It turned out that the tooth belonged to an extinct relative of swine. Pigs and people have much in common, especially their digestive systems, which explains why the two have formed such an enduring, albeit fraught, relationship.

and people have much in common. The two species have similar digestive systems, from teeth to stomach to intestines, because they have similar diets. Both are omnivores who thrive on meat, nuts, roots, and seeds. And because pigs and people eat the same foods, they evolved to form a symbiotic connection—a bond so tight that 10,000 years later, it remains unbroken.

A giant meteor smashed into Earth about 65 million years ago. The meteor kicked up dust, the dust changed the climate, and the new climate killed off the dinosaurs. Onto the freshly cleared playing field stepped the mammals. These warm-blooded, lactating creatures had first emerged about the same time as dinosaurs, but for millions of years they had remained minor players, mouse-sized beasts scurrying about the forest floor. When the dinosaurs died, mammals rose to the occasion, growing larger and filling just about every available niche.

About 10 million years after the meteor struck, the first hoofed mammals, or ungulates, appeared. One order of ungulates, called *Perissodactyla*, includes just a handful of living species, such as horses, rhinoceroses, and tapirs. The other order, *Artiodactyla*, is much larger and includes pigs, cows, goats, sheep, camels, llamas, giraffes, deer, antelopes, camels, hippopotamuses, bison, and water buffalos. Both orders of ungulates might be called tiptoers. Their hooves are actually outsized toenails, and they walk like ballerinas *en pointe*. The arrangement of those toes divides artiodactyls from perissodactyls. "Perissodactyl" means "odd-toed": the foot's axis cuts through the center of the middle digit, and the animals walk either on three toes, like rhinos and tapirs, or just one, like horses, zebras, and donkeys. "Artiodactyl" means "even-toed:" the first digit (the thumb or big toe) is absent, and the feet are symmetrical, with the axis running between the third and fourth digits (the equivalent of the human middle and ring fingers). As they evolved to move more quickly, their outer digits shrank or disappeared, letting the animals run on just the middle two digits. Thus they appear to have a single hoof split down the center, what the King James Bible describes as the "cloven foot."

Subtropical forest dominated the Northern Hemisphere when ungulates first evolved, and they dined on tender leaves, seeds, and fruits. About 20 million years ago, the climate became cooler and drier, forests disappeared, and grasses spread over millions of square miles. Deprived of their old forest habitats, some hoofed animals adapted to the new circumstances. In forests, hiding was the key strategy to avoid predators, but grasslands offered little cover, so the bodies of many ungulates evolved: their eyes shifted farther to the back on their heads and became larger, allowing them to see predators more easily. And they became cursorial, or primed for running: free-swinging knee joints and longer, stiffer leg bones gave these animals enough speed to outrun a big cat.

These graceful savannah creatures ate grass, which is rather like chewing on sandpaper. Grass cells contain minute glassy particles, and the blades often pick up an additional coating of grit from the dirt below. If humans tried to eat grass, they would wear their teeth down to the gums. To deal with the new diet, many ungulates evolved teeth that grow constantly, rather like mechanical pencils, with new material emerging from the gums as the top wears away.

Those newfangled teeth solved only part of the problem. Grass is heavy on cellulose, which consists of simple sugars bound together so tightly that no enzyme produced by mammals can break them apart. That process requires the assistance of bacteria, which live in the gut and break down cellulose through fermentation, making the sugars and other nutrients available to the animals. Cows and sheep—along with giraffes and deer—bite off and swallow large amounts of grass without chewing it, and it passes into the rumen, or first stomach, where bacteria begin to digest the cellulose. Then, when the animal is resting, it regurgitates the food and "cheweth the cud" (as Leviticus tells us) before swallowing it again and allowing the grass to pass all the way through the alimentary canal.

These developments in many ungulates—rapid running, sharp eyesight, and the ability to eat grass—led to the extraordinary success of hoofed animals. Artiodactyls demonstrate especially beautiful and astonishing specialization: the gazelle bounding across the savannah, the giraffe grazing the tops of trees, the mountain goat scaling a vertical cliff, the powerful bison roaming the grasslands of America.

And then there are pigs. The pig and its close cousin, the peccary, are the odd men out, artiodactyls that didn't become ruminants. While their cousins signed up for the evolutionary fast track, moved to a new territory, and accomplished great things, the pig stayed at home in the forest. And that has been the key to its success.

Being a generalist has advantages. The very earliest mammals ate insects and had teeth like those of reptiles, simple spears for holding bugs until they could be maneuvered into position to be swallowed whole. The mammal jaw then evolved to become stronger and more dexterous. It could move side to side as well as up and down, allowing animals to chew a variety of foods and thereby get the process of digestion started earlier. Eventually, mammals developed a full complement of incisors, canines, premolars, and molars adapted to shearing, slicing, grinding, puncturing, and crushing. The more kinds of teeth you have, the more kinds of food you can eat.

Whereas other hoofed mammals gave up those generalized skills, pigs stayed true to the forest-dwelling first mammals. They kept shorter limbs, the better to scoot through the brush. Since pigs lived in dense thickets, they didn't need good eyesight, so their eyes remained small. Since good hearing was an advantage, their ears remained large, and they learned to communicate through a wide variety of grunts and squeals. Adapted to moist, shady environments, they have few sweat glands and cool themselves with a wallow in the mud. Pigs are not good at standing in a field in the hot sun. That is a job for cows.

The pig's most specialized and distinctive feature, the snout, allows it to take advantage of the forest environment. In humans, the somatosensory cortex—the part of the brain responsible for sensation—is wired primarily to the hands. In pigs, nearly all the touch-sensitive nerves terminate in the nose. It's best to think of a pig snout not as a nose at all but as something like an elephant's trunk, a miraculous fifth limb that allows

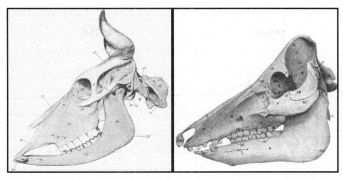

Like humans, pigs (right) have molars, premolars, canines, and incisors that allow them to slice, rip, and grind a wide variety of foods, from tender plants to the tough flesh of large animals. By contrast, cows (left) have incisors and molars, suitable only for cropping and chewing grass and leaves.

the pig to react to its world in ways unknown to other hoofed mammals. A tough cartilage nasal disk allows the pig to plow into rock-hard ground, while a fine mesh of snout muscles lets the pig make delicate rooting motions without moving its head. Other muscles clamp the nostrils shut to keep out dirt while still allowing puffs of air to enter, so that the pig's exquisite sense of smell can determine whether a hard, round object is a rock to be nudged aside or a nut to be cracked open. Despite constant rough use, the snout remains, in the words of one pig observer, "art-gum-eraser tender," as sensitive and finely tuned as a safe-cracker's fingertips.

Like the multichambered stomach that allowed other artiodactyls to eat grass, the pig's nose marked an immense evolutionary leap. The snout opened the underground realm to the pig, vastly increasing the amount of food available to it. Subterranean roots and tubers were relatively unaffected by forest fires, drought, or overgrazing by ruminants, giving pigs an advantage over other ungulates during hard times.

Those roots suited the pig's digestive system well. Compared to those of ruminants, the pig's intestines had a limited array of bacteria to ferment plant matter, so grass and tough leaves were off the menu. Pigs instead ate bulbs, tubers, seeds, nuts, and fruits, which are packed with easily digestible simple sugars and proteins. They also ate tender plants, fungi, insects, worms, grubs, snakes, lizards, ground-nesting birds, small mammals, fish, clams, and carrion. This diet required an anatomy rather different from that of cows. Rather than a complex gut and single-purpose teeth, pigs went with a simple gut and multipurpose teeth.

Because the pig evolved into a dietary generalist, its digestive system greatly resembles that of a person. Herbivores have turned their guts into giant fermentation tanks to allow them to eat leaves and grass, and carnivores such as lions have powerfully muscled stomachs that can churn up large chunks of meat into small bits of usable protein. But the pig developed no such skills. Its simple stomach chops up proteins, its small intestine absorbs sugars and other nutrients, and its colon sucks up water and does its best to ferment any plant material. Roughly speaking, that's the same gut design found in humans, chimpanzees, and orangutans, not to mention a fair number of lower primates. Like pigs, all of these animals eat nuts, fruits, tender leaves, insects, and meat. This type of gut is remarkable for its lack of specialization: it can adapt to nearly any circumstances.

An expansive menu requires enhanced intelligence. One scientist who studies pig cognition complained that no one was surprised by his findings: "I would recommend that somebody study sheep or goats rather than pigs, so that people would be suitably impressed to find out your animal is clever." The flaw in that plan is that sheep and goats aren't terribly clever; animals whose only dietary task is to spot something green and start chewing have little need of higher mental powers. Omnivores, by contrast, face difficult choices. They must be open to novel foods because individuals hardwired to discover new sources of nutrition tend to thrive and pass on their genes. But indiscriminate snacking poses dangers: pigs that eat toxic mushrooms don't leave many offspring. We might think of the pig as a judicious risk taker, open to the new but capable of assessing potential threats. In that quality, pigs are much like people.

Many developments set humans apart from their more apelike ancestors. The first important shift was bipedalism: 4 million years ago, our ancestors developed a skeleton adapted to walking upright rather than climbing trees. They started using stone tools 3 million years ago. Along the way they developed more acute vision, lost much of their body hair, and evolved more efficient sweat glands. The most important changes, however, involved the brain and the gut.

Digesting and thinking are the most energy-intensive processes in animal physiology. According to what scientists call the "expensive-tissue hypothesis," before an animal can develop a big brain, it must first lose its large gut, because having both would exact an enormous cost in calories. The only way to shrink the gut is to subsist on higher-quality food—not grass but nuts, fats, and meats. In the human lineage the crucial shift took place about 1.8 million years ago, when *Homo erectus* appeared. Compared to their most immediate predecessors, these human ancestors had bigger skulls, smaller teeth, and a smaller rib cage and pelvis—the last two providing evidence of a smaller gut. Most likely, these changes were linked not simply to eating nutrient-rich foods but to cooking them, which made digestion even more efficient. By cooking their meats and roots, our ancestors freed up energy that otherwise would have gone to digestion, allowing it to be redirected to the growth of a bigger brain.

Scientists developed the expensive-tissue hypothesis to explain human evolution, but it might also account for why pigs are smarter than cows. The enormous guts of ruminants, required to ferment grass into digestible sugars, spare few calories for the brain. Pigs, with simple guts and calorie-intensive diets, can devote more metabolic energy to thinking.

The ancient fossil beds of the world are littered with the skeletons of barrel-bodied, short-limbed, piglike creatures. Only two families—pigs (*Suidae*) and peccaries (*Tayassuidae*)—survive today. Their evolutionary lines split more than 35 million years ago in Asia. The *Suidae* line stayed in the Old World, evolving into the wild boar of Eurasia (ancestor of the domestic pig), the warthog of Africa, the babirusa or "pig deer" of Indonesia, and a dozen or so other species. Members of the *Tayassuidae* line live only in the Americas. The peccary's head is a bit shorter than the pig's, its tusks point down rather than up, and it runs faster. Otherwise pigs and peccaries are remarkably alike. Thick body, small eyes, probing snout, multipurpose teeth—all are adapted to omnivorous life in the bush.

Pigs are often described as the most "primitive" of the artiodactyls. In a sense, they are: they have the full range of teeth and the simple guts of early mammals. In this sense, however, people too are primitive. Sometimes the simplest tools are the best. An ecological niche, at root, is just a source of energy, a place in the world where a plant or animal can find enough food to sustain itself. Pigs and people, both expert at evaluating new foods, quickly colonized every remotely viable niche.

Some 31 million years after the pig's *Suidae* ancestors split off from their *Tayassuidae* cousins, the family divided again, forming a new species. The Eurasian wild boar—classified as *Sus scrofa*, Latin for "breeding sow"—first evolved in Southeast Asia 4 million years ago and then radiated throughout the continent. Today the natural range of the Eurasian wild boar stretches from northeastern Europe to Southeast Asia, from 13,000-foot mountains to swamps to near desert. Pigs, like people, are everywhere, and for many of the same reasons: they have clever brains, omnivorous appetites, and general-purpose teeth and guts.

The Eurasian wild boar did very well in the wild, but it thrived as never before when, about 10,000 years ago, certain members of the species took one further evolutionary step: they gave up their independent ways, moved into town, and domesticated themselves. Pigs and people threw in their lots together, and that proved a wise evolutionary strategy for both species.

TWO

OUT OF
THE WILD

In 1989 archeologists discovered a tiny ancient village in the
foothills of the Taurus Mountains and began digging with
haste: the Turkish government was building a dam on the nearby
Batman River, a tributary of the Tigris, and the area would soon
be inundated. The site, known as Hallan Cemi, dated to about
11,000 years ago. Most of the buildings the archaeologists excavated were tiny round huts of wattle and daub, just six feet
in diameter, though there was one larger ceremonial building
with a cattle skull mounted on the wall. Among the debris were
beautifully carved sandstone bowls, decorated grinding stones,
and obsidian tools.

The villagers of Hallan Cemi had formed a complex society with a rich cultural life, but they were hunter-gatherers, not
farmers. They ate wild lentils, bitter vetch, almonds, plums, and

pistachios. They hunted sheep, goats, deer, and pigs. The first three were unquestionably wild animals. The site's pig bones, however, told a more complicated story.

Nearly half the pigs eaten at Hallan Cemi were killed at less than a year old, a profile very different from the broad age range found in hunted animals. The bones also were overwhelmingly male, suggesting that the villagers had spared females to serve as breeding stock. During the time that Hallan Cemi was occupied, moreover, deer bones became less common while pig bones increased in number. Both deer and wild pigs are forest-dwelling creatures, so if the pig bones at Hallan Cemi came from wild animals, their numbers should have declined along with those of the deer. The fact that they did not makes it likely that the destruction of forests killed off the deer, while pigs found a new habitat, living alongside humans in the village.

The changes at Hallan Cemi form one small part of the most significant event in human history: the invention of agriculture. This era, known as the Neolithic or New Stone Age, arose independently, and at very roughly the same time—10,000 to 5,000 years ago—in Asia, the Near East, the Americas, and Africa. Given how widespread it was, the rise of farming was likely triggered by a single phenomenon: the end of the ice ages and the warming of the climate. But many mysteries remain. Why would anyone choose the rigors of agriculture over the comparatively easy life of hunting and gathering? Why did prehistoric humans domesticate some animals and plants but not others? And how did the process of domestication take place?

Hallan Cemi lets us begin to answer some of those questions, at least with regard to pigs. The archaeological record suggests that *Sus scrofa* took up life in villages because, as omnivores, they were biologically equipped to eat the garbage generated by newly sedentary humans. Most hunter-gatherers, after all, moved on to a fresh campsite before the stench grew too foul. In the first permanent settlements, pigs acted as sanitation service and food source, transforming butchery waste, rancid almonds, and moldy wheat into meat. Pigs proved useful to people—but that doesn't mean the relationship developed solely because of human intention.

It's commonly held that humans domesticated pigs, but that's not quite right. Wild boars, in adapting to the new niche created by human settlements, evolved in ways that made them capable of living in close proximity to people. Pigs, in other words, domesticated themselves.

A natomically modern humans, *Homo sapiens*, arose about 250,000 years ago in Africa. Between 60,000 and 130,000 years ago—there is much disagreement on timing—they crossed the Red Sea, entered the Arabian Peninsula, and from there spread all over the world. That means the Eurasian wild boar and modern humans first crossed paths in the Near East, that confluence of continents in the eastern Mediterranean.

The eastern Mediterranean is generally an arid land, but coastal marshes and the riverbanks provided habitat for pigs. The amount of good habitat available varied considerably, because over the last 2 million years the Earth's climate has shifted from warm and wet to cold and dry every 100,000 years or so. About 20,000 years ago—some 9,000 years before people settled at Hallan Cemi—the Earth was in the grip of the last great ice age. Rivers dried up, and ice locked up so much of the Earth's water that the coast of the Mediterranean Sea was five or ten miles from its current location. Steppes covered with grass and scrubby brush dominated the landscape, and forests survived only in the uplands. Humans in the region, like those

The wild boar emerged in Southeast Asia about 4 million years ago and then colonized nearly every corner of Eurasia, from mountain to desert to jungle. When the first groups of *Homo sapiens* ventured out of Africa into the Near East some 60,000 years ago, *Sus scrofa* was already there waiting for them.

everywhere else in the world, lived as hunter-gatherers. They collected wild grains, beans, and nuts and used stone-pointed arrows and spears to kill game.

When the weather warmed about 15,000 years ago, the people of the Near East found themselves living in a world where food was abundant. They gathered nuts and fruit from forests of oak, pistachio, almond, and pear trees and harvested grains and legumes from huge stands of wheat, barley, rye,

genetic mutation—the seeds remained high above the ground, unable to germinate, directed down an evolutionary dead end. But then humans came along. On a few occasions, they would have come across a stand of wild cereal after it had ripened, and the plants with this mutation would have made up a disproportionate percentage of the harvest. People would have kept these seeds to replant, thereby perpetuating the mutations. This reversed the course of evolution: an undesirable quality in nature became indispensable to human culture. Soon humans selected for other desirable qualities in the plants, such as increased seed sizes and thinner seed coats. Thanks to these first experiments in plant breeding, people enjoyed more bountiful crops that were easier to harvest, thresh, mill, and eat.

Mutated wheat and barley was waiting for humans, ready to be plucked and planted, but animals were different. People could kill and eat them, but they couldn't take wild beasts back to the village and commence a captive breeding program. Domesticating animals entailed a more complex process.

Throughout history people have tamed all sorts of wild animals—kangaroos, monkeys, cassowaries, cheetahs, moose, giraffes, and brown bears—apparently for the simple pleasure of keeping pets. But when those individual animals die, their tameness dies with them. Domestication, by contrast, is an evolutionary process in which a species adapts to living alongside humans. Whereas taming involves individual animals, domestication involves populations. Pet keeping might be thought of as an initial evaluation to see if the animal might work out as a permanent employee: many animals were interviewed, but few were hired.

Those animals that made the cut—large domestic beasts like cows, pig, sheep, goats, horses, camels, and llamas—have certain things in common. They eat widely available foods such as grasses and leaves (unlike, for instance, koalas, which require specific types of eucalyptus leaves). They reproduce fairly quickly (unlike elephants, which gestate for almost two years). They respond to threats by bunching up rather than by scattering (as white-tailed deer do). They aren't too aggressive (like African buffalo) or agile (like fence-leaping gazelles). Perhaps most importantly, they live in groups structured by a dominance hierarchy, a trait that allows humans to step in at the top position and assert control of the herd.

This pattern helps explain why certain animals became domestic while others didn't. But it doesn't allow us to peer back into history to see precisely how these animals first made the transition from a wild to a domestic state. In order to understand that, we must look at individual cases of domestication.

The domestic goat, which likely emerged from the shifting hunting tactics of humans, offers a relatively clear-cut case. All of the goat breeds that exist today are descended from the wild Bezoar goat, which lives in high mountains stretching from Turkey into Pakistan. Archaeologists in the region have uncovered ancient heaps of goat bones in caves occupied by humans during a period stretching from 45,000 years ago to about 9,000 years ago. For nearly all of that period, most of the bones came from adult animals, both males and females. Then, starting about 9,000 years ago, the bones of subadult males begin to predominate. After that time, very few male goats lived to adulthood. By contrast, almost all of the females did.

In the earlier periods, it seems, the hunters focused on immediate returns, killing adults because those animals had the most meat on their bones. The later pattern—killing males at younger ages while allowing females to grow older—reflected what we might call a sustainable harvesting strategy, intended to ensure that animals would be available for hunting in the long term. The shift most likely reflected dwindling resources. At this time, roughly 12,000 years ago, the people living in the mountains from present-day Turkey to Pakistan underwent the same climate-induced stresses as those at lower altitudes. The weather warmed, and human population boomed, increasing the hunting pressure on local animal populations; this, combined with a cooling climate, caused the herds of wild goats to dwindle. The hunters changed their hunting strategies to meet the new conditions. They killed young males—only a few were needed for breeding—and protected females, who gave birth to kids and thereby sustained the herd's population.

What started as a hunting strategy became, over hundreds or thousands of years, a new relationship with goats. Hunters followed the herds through the hills, closely observing them, learning their behavior, and selectively culling them to maximize docility and productivity. In time, hunters gave way to shepherds, who controlled their precious livestock with fences and herding dogs.

A similar domestication process is likely for sheep and cows—but not for swine. Omnivorous and intelligent, pigs demand a different theory of domestication, one that accounts for the ways they manipulated people.

Wild boars, *Sus scrofa*, became domestic pigs, *Sus scrofa domesticus*, several times, independently. It happened once along the Yellow River in China and perhaps a second time along the Yangtze sometime before 6000 BC. It's also likely that pigs were domesticated in India and Southeast Asia. The occasion we know most about happened in the Near East, starting at Hallan Cemi, roughly 11,000 years ago.

When something happens a half dozen times or more, it starts to take on an air of inevitability, as if the partnership between pigs and people had been rendered necessary by the nature of each. The habits of wild pigs had not changed much in the few million years before domestication. The same cannot be said of people: after the last great ice age, they had stopped wandering in search of food and settled down into villages. Not long afterward, pigs joined them.

The process of pig domestication didn't happen as it did with goats because wild boars aren't much like wild goats. Pigs group themselves not in large herds but in small maternal family groups, known as sounders, averaging a dozen or so individuals. And they live not on open hillsides but in the forest. To hunt them, humans likely sat quietly and waited for one to wander down the path. Boars couldn't be followed, observed, and selectively killed. Boar hunting could not have turned into boar herding.

A different mechanism was at play, and the story of how wolves turned into dogs might provide the key to understanding it. Dogs were the first domestic animals, emerging 15,000 or 30,000 years ago, long before the invention of farming. According to an earlier theory of dog domestication, our ancestors snatched wolf pups from the den, tamed them, and carefully bred a new species to serve as companions, sentinels, and hunting partners. The first stage, adopting pups, surely happened, given what we know about pet keeping. But the second stage, transforming these captured pups into dogs, poses problems. Modern experiments show that wolves hand-raised from the age of eight days become relatively tame but are most certainly not dogs: they bite their handlers, cannot be trained to sit or stay, and try to run off as soon as they reach sexual maturity.

More likely, wolves entered the school of domestication not as pets but as scavengers. Wolves trailed bands of hunter-gatherers, devouring the remnants of the hunt. When people settled into permanent villages, they produced even more materials for wolves: not only animal scraps but also burnt food, rotten fruits, and spoiled grains and nuts. A genetic mutation that allowed wolves to better digest starch may have played a role in their domestication. Some of the waste would have been cooked grains and meat, which offered further advantages: the process of cooking renders more of the calories in food available, offering animals who could exploit this resource a particularly rich source of energy. Accessing this food posed risks, because humans tend to be hostile to large wild animals wandering among them. Biologists judge an animal's wariness in terms of "flight distance"—how far an animal will run away if a human approaches. Flight distance is, in part, genetically determined, and variation exists within populations. Those wolves with shorter flight distances—that is, with a greater tolerance for proximity to humans—claimed more food and reproduced more successfully.

Boars likely underwent a similar domestication process. Wild boars, unlike wolves, did not need to await a genetic mutation to be able to feed on cereal grains: they had always enjoyed the ability to eat starches, along with just about any other substance humans ate. Like dogs, they would have been attracted to the new source of food found in human garbage heaps. Also like dogs, they would have had to undergo an evolutionary adaptation—becoming less fearful of humans—in order to exploit it.

The selective forces of this new ecological niche favored individuals that were bold (not prone to flee at the sight of humans) but not aggressive (because humans would kill those that posed a threat). The wild animals began to separate into two populations: one tolerated human presence; the other did not.

The boars and wolves most adept at exploiting this food source bred themselves into the new, human-compatible species that we now know as pigs and dogs.

With dogs, this theory of domestication is largely speculative, based on what we know of dog and wolf behavior. With pigs, however, we can test the theory against archaeological evidence. Ancient dog bones are not uncommon, but their numbers pale in comparison to those of pigs. For every dog Neolithic people kept as a sentry, hunting companion, or pet, they kept dozens of pigs as a food source. People ate pork in quantity and tossed the bones into piles, where they remained buried for millennia, awaiting analysis.

The earliest evidence of pig domestication comes from Hallan Cemi, the site now buried under a lake in Turkey. There, villagers ate mostly young male pigs, and pig bones became more common even as the forest dwindled, suggesting the pigs lived as scavengers within the village rather than as wild beasts roaming the nearby woods. All of the pigs at Hallan Cemi, however, were wild boars: their bones show none of the morphological changes, such as shorter snouts, that indicate domestication. This may have been a simple matter of time: Hallan Cemi was occupied for just four hundred years, and the domestication process may have been only getting started when the village was abandoned.

A longer run of evidence comes from a site known as Cayönü Tepesi, not far from Hallan Cemi along the Boğazçay River, another tributary of the Tigris. Settlement at Cayönü started slightly later than at Hallan Cemi, around 8500 BC, and the site was continuously occupied for nearly 2,000 years. And over that span, the pig bones change from wild to domestic.

Because domestication is an evolutionary process, there is no clear threshold when a group of animals stops being wild. It's

Neolithic people drew this image of a wild boar hunt in what is now Turkey in about 7000 BC, roughly the same time and place that the animals first entered into partnership with people. Pigs domesticated themselves, lingering near human settlements to scavenge food and gradually evolving into domestic creatures. (Courtesy Omar Hoftun, Creative Commons Attribution–Share Alike)

better to think of the human relationship with such animals as a continuum: it starts with hunting, moves on to various forms of human management of wild populations, and continues to full domestication.* Evidence of this process might start with a shift in the demographics of the animals killed—such as the goats in the Zagros or the pigs at Hallan Cemi—and gradually proceed to a change in morphology. This is exactly what we see at Cayönü Tepesi.

As time progressed at Cayönü, the village's residents killed pigs at progressively younger ages, suggesting human control similar to that at Hallan Cemi. Hundreds of years later,

---

*As America's feral swine problem indicates, this process also works in reverse.

evidence began to appear in the structure of the bones as well. Compared to their wild ancestors, domestic pigs have shorter snouts, smaller brains, and more crowded teeth. These changes are most likely a side effect of domestication. Domestic animals are selected for docility, which is a juvenile trait—young pigs are less aggressive than adults—and the genes for docility come packaged with other juvenile traits, such as a shorter snout and smaller teeth. Just as domestic pigs preserve their juvenile docility into adulthood, they hold on to a more juvenile skull shape too. Scientists know that when they find adult facial bones of a certain shape, they are looking at a domestic pig.

The changing shapes of the bones at Cayönü Tepesi indicate a gradual process of domestication. The archaeological record reveals that, over the 2,000-year period of settlement at the site, wild boars lived among people, gradually evolving into domestic pigs.

The same continuum from wild to domestic characterizes the archaeological record of other domestic animals, but it's worth noting a crucial difference. Goats and sheep became domestic through their role as prey for human hunters: people first killed wild animals, then managed wild herds, and finally managed domestic herds. Pigs became domestic through their relationship not with humans as hunters but rather with humans as villagers. People tracked down goats, but pigs tracked down people. Once domesticated, goats retained their original habitat, the scrubby hills and grassland outside town, whereas pigs took up residence right alongside humans. From the start, it was a more intimate relationship, involving everyone who lived in town rather than just herders assigned to the task.

Pigs, moreover, had a job to do beyond providing meat. They cleaned up the waste that accumulated in each village they occupied: dead animals, rotten food, and human feces. The villagers could not have understood that this was a useful public health measure, but they would have been happy to be rid of the stink.

Pigs possessed alchemical powers, transforming garbage into food. At first, this must have seemed like a godsend. In time, however, people came to despise the pig for doing precisely the job it had evolved to do.

THREE

"THE PIG IS IMPURE"

Built in about 2550 BC, the Great Pyramid of Giza stands 455 feet tall and comprises some 2.3 million blocks of stone weighing about 13 billion pounds in aggregate. Archaeologists still argue over whether those stones were moved into place using levers, sledges, or oil-slicked ramps. Whatever the technical method, building the pyramids involved a feat of social engineering just as impressive as the mechanical: Egyptian authorities had to feed a workforce of thousands of people for decades at a time.

The builders of the Great Pyramid called upon the resources of the entire Nile Valley to support this effort. The royal house sent orders to the heads of villages, who in turn sent men to the Giza site, along with grain and livestock to feed them. Workers drank beer, a muddy beverage fermented from grain and consumed more for nutrition than for pleasure. They ate heavy loaves of wheat and barley, supplemented with beef, mutton, and goat. One archaeologist analyzed some 300,000 bones at the pyramid complex and found that nearly all the animals eaten were young and male. This proved that Giza was a provisioned site, with animals raised elsewhere and the juvenile males—not needed for breeding—marched to slaughter at the pyramids.

One village that provided livestock was Kom el-Hisn, located in the Nile delta about seventy-five miles downriver from the temple complex. Villagers at Kom el-Hisn raised cattle but ate very little beef: only the bones of worn-out breeding cows and sick calves have been uncovered there. Instead, the villagers ate pork: for every four cattle bones archaeologists unearthed at Kom el-Hisn, they found one hundred pig bones. It seems that the residents kept herds of pigs that foraged in the Nile delta marshes and scavenged trash on streets. Although Egypt's rulers demanded cattle from Kom el-Hisn, along with goats and sheep from other settlements, the villagers' pigs were spared.

The reasons for this had to do with climate and biology. Animals destined for Giza had to walk hundreds of miles through an arid landscape, feeding on grass and leaves along the way. Well suited for such a journey, cows, goats, and sheep were herded to Giza by the thousands. Pigs, however, would not have found the food or shade they needed along the way. The state couldn't move pigs around, so it ignored them.

This pattern appeared throughout the Near East: officials developed complex food-provisioning systems that depended on the long-distance movement of cows, sheep, and goats. Pigs didn't fit into such schemes. But despite—or perhaps because of—their lack of usefulness to bureaucrats, pigs didn't disappear. Instead, they stuck to their original role as scavengers. People on the fringes of society with little or no access to state-supplied food embraced them as a source of meat. Priests and

bureaucrats, who dined on lamb and beef, came to despise pigs.
Only the poor ate pork.

For its first 4,000 years, agriculture remained a modest affair. The farmers of the Near East lived in mud-brick huts
in villages ranging in size from a few dozen to a few thousand
people—places like Kom el-Hisn, Cayönü Tepesi, and Hallan
Cemi, which did not change much from one century to the next.
     The pace of change picked up about 5500 BC. That's when
people in Mesopotamia—the lands around the Tigris and Eu-
phrates Rivers in present-day Syria and Iraq—developed irrigation agriculture. A thousand years after that, the plow appeared.
The first true cities, with tens of thousands of residents and
complex social organizations, appeared about 3500 BC. The
Mesopotamians invented writing—first pictographs and later
the more abstract cuneiform—and built the first monumental
temples, called ziggurats, to worship their gods. Across the Red
Sea, Egypt got a slightly later start but achieved more lasting
success. By about 3000 BC, Egyptian rulers had unified a ribbon
of land stretching for six hundred miles along the Nile. Scribes
created a hieroglyphic writing system about this same time, and
laborers were put to work on pyramids.
     Culture depends on agriculture, and in Egypt and Mesopo-
tamia the two flourished together. Both empires emerged from
desert landscapes along rivers. No one had settled these areas
earlier because there wasn't enough rain for farming, but irrigation allowed farmers to exploit the rich soil deposited by
seasonal floods. That soil produced crops in great abundance,
which meant some members of society could give up agricultural work and devote themselves to making crafts (pottery,
baskets, bricks, tools, weapons), building temples, keeping

records, fighting battles, and serving the gods. "A human being is primarily a bag for putting food into," George Orwell once wrote. "The other functions and faculties may be more godlike, but in point of time they come afterwards." Only when farmers grew enough food to fill the bellies of bureaucrats, priests, and soldiers could these elites go about the business of creating what we call civilization.

Mesopotamia and Egypt built centralized economies and strictly controlled the distribution of grains, dairy products, and meat to the population. The city of Puzrish-Dagan, for example, served as an administrative center for Mesopotamia's Third Dynasty of Ur, which lasted from 2112 to 2004 BC. Surviving records show that the ruling dynasty requisitioned tens of thousands of animals from outlying areas. One archeologist tabulated the records from this economic center, tracing the flow of more than 10,000 animals that arrived from the provinces and were then distributed throughout the urban center. The temple claimed lambs and kids, and soldiers ate cattle and older sheep. The records make no mention of pigs. As in Egypt, they existed but held no interest to the state.

Villagers in both Mesopotamia and Egypt kept pigs purely on their own initiative. Throughout the Near East, pigs could be found wherever there was water. Towns near natural pig habitats—along the Jordan River, for instance—kept the most pigs because the animals could supplement urban scavenging with foraging in the woods and marshes. Towns in drier areas kept fewer pigs. Nomadic pastoralists, on the move for much of the year, kept none. Archaeologists have plotted on maps the areas that received enough rainfall to allow farming without irrigation. All villages within those areas showed evidence of pig remains. In other words, if it was biologically possible to raise pigs, people raised pigs.

There were variations within this broad pattern. At Tell Halif, a small site on the edge of the Negev desert in what is now southern Israel, the archaeological record shows dramatic swings in the reliance on pork: pigs account for more than 20 percent of animal bones in garbage heaps dating to 3000 BC. That figure plunges to less than 5 percent five hundred years later, rises again to 20 percent by 1500 BC, and finally drops once more to less than 5 percent by 1000 BC. Changes in rainfall levels cannot explain those swings. It seems that the true reason was political: periods of highest pig use correspond with times of weakest state control. Halif was located along a major trade route; when the political situation was stable, the town likely became integrated within a regional economy, and a steady supply of sheep and goats flowed through. When the ruling dynasties descended into chaos—as they did rather frequently—the town had to fend for itself. That's when the villagers turned to pigs.

The rise of strong states discouraged pig raising in another way as well: by changing the landscape. As populations grew, they put increased pressure on the land. Farmers felled oaks to make way for olive groves and drained marshes to plant crops. The land, often poorly managed, deteriorated from forest to cropland to pastureland to desert, with each successive stage providing less habitat for pigs. By the time desert scrub prevailed, only sheep and goats could survive. As pigs lost habitat, they likely began to raid crops in the field, threatening the food supply and thereby earning a spot on the state's hit list.

Pigs didn't fit into the new political and agricultural order. As time marched on, they began to disappear. At many archaeological sites, pig bones remain common up through about 2000 BC, then dwindle away. A thousand years later, few people raised pigs in any quantity.

In a few spots, however, pigs persisted. They remained important for sites like Tell Halif that were on the margins of empire, far from the urban centers. And pigs became crucial to the marginal people living within those urban centers. Careful sifting of debris from streets has turned up shed milk teeth—baby teeth—of piglets, evidence that pigs were living and breeding among the homes of the world's first great cities. But not everyone in those cities partook in equal measures. Archaeologists tend to find pig bones in the areas of cities where the common people lived. In elite areas, they find more cattle and sheep bones.

Some of the most compelling evidence of this pattern comes from the temple complex at Giza. At the official barracks, temple laborers ate provisioned beef driven there from far-flung villages. Nearby, however, another settlement grew up. This neighborhood, haphazardly constructed, most likely housed those who provided services to temple workers and bureaucrats—grinding wheat, baking bread, brewing beer. These people were not part of the official workforce and therefore did not receive food directly from the rulers. Instead they hunted, foraged, and traded for their food, or they raised it themselves. And what they raised was pigs. Although absent from the residences of official workers, pigs are common in this self-supporting area. Pork offered these common people what we would call food security: a source of meat under their own control.

Poor people ate pork because it was the only meat they had. The elite refrained from eating it because they had access to other sources of meat. In time, though, the ruling classes began to actively avoid pork. The Greek historian Herodotus, in the fifth century BC, reported that an upper-class Egyptian man, after accidentally brushing against a pig, rushed into the Nile fully clothed to cleanse himself.

By the start of the Iron Age, about 1200 BC, elites in the Near East had begun to see pigs as polluting, a view that arose in part from the habits of urban pigs. Though cities had grown large, sanitation systems had not kept pace. Residents threw garbage into the streets or piled it in heaps outside their doors. This waste included spoiled food, dead animals, and human excrement. Information about ancient sewage disposal is scant; one of the few references is found in Jewish scripture. "You shall have a stick," Moses tells his people in Deuteronomy, "and when you sit down outside, you shall dig a hole with it, and turn back and cover up your excrement. Because the Lord your God walks in the midst of your camp, . . . therefore your camp must be holy, that he may not see anything indecent among you, and turn away from you."

Evidence suggests that the Lord God saw quite a few indecencies among the Israelites and their neighbors. Sewer systems didn't exist. A few elite homes and temples had pit latrines, but mostly people practiced what today is known as open defecation: they relieved themselves in fields or streets, and they didn't bring a stick. This is where pigs enter the picture.

Pigs eat shit. In many villages around the world today, pigs linger around peoples' usual defecation spots awaiting a meal. Some English pigs in the eighteenth and nineteenth centuries had the same habit. In China, archaeologists discovered a terra cotta sculpture, dating to about 200 AD, showing a pig in a sty, with a round, roofed building just above it. The structure was originally identified as a grain silo for storing pig feed, but the model in fact depicted a combination pigsty-outhouse: people sat on an elevated perch and made deposits to the hungry pig

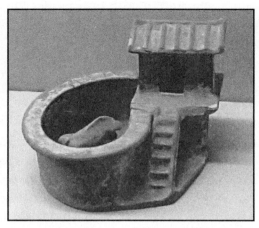

This Chinese sculpture, dating to about 200 AD, depicts an outhouse perched over a pigsty. All over the world pigs ate human waste, carrion, and rotting garbage, a habit that made them quite useful—they cleaned the streets and transformed filth into meat—but sometimes turned them into pariahs. (Courtesy John Hill, Creative Commons Attribution–Share Alike)

below. The practice was widespread—the same Chinese character designates both "pigsty" and "outhouse"—and has survived into the present on Korea's Cheju Island. In the 1960s more than 90 percent of farmers on the island used a pigsty-privy in their subsistence-farming regimen, and they insisted the arrangement produced the sweetest pork in the world.

The pigsty-privy apparently did not exist in the ancient Near East, but pigs discovered this food source on their own. Tapeworm eggs have been found in fossilized pig feces from ancient Egypt. Since these eggs are produced only by adult tapeworms living in human guts, it appears that human feces formed part of Egyptian pigs' rations. In Aristophanes' play

Peace, dating to the fifth century BC, a character notes that a "pig or a dog will . . . pounce upon our excrement."

This particular dining habit did not improve the pig's reputation. Just as troubling was the pig's taste for carrion, including human corpses when available. Eating human flesh and eating excrement are nearly universal human taboos, and eating animals that eat those substances carried a transitive taint. "The pig is impure," a Babylonian text asserted, because it "makes the streets stink . . . [and] besmirches the houses." An Assyrian text from the 670s BC contains these curses: "May dogs and swine eat your flesh," and "May dogs and swine drag your corpses to and fro on the squares of Ashur."

Dogs and pigs had first domesticated themselves by scavenging human waste, but now that role made them pariahs. Filthy animals offended the gods and therefore were excluded from holy places. The people of the Near East practiced many different religions, but all agreed that the key sacrificial animals were sheep, goats, and cattle and that pigs were unclean. In Mesopotamia and Egypt, pigs never appear in religious art. The Harris Papyrus, which describes religious offerings made by King Ramses III, includes a detailed list of every desirable item to be found in Egypt and the lands it had conquered, including plants, fruits, spices, minerals, and meat. Pork does not appear on the list. "The pig is not fit for a temple," a Babylonian text reads, because it is "an offense to all the gods." A Hittite text declares, "Neither pig nor dog is ever to cross the threshold" of a temple. If anyone served the gods from a dish contaminated by pigs or dogs, "to that one will the gods give excrement and urine to eat and drink."

Many people, for many different reasons, rejected pork in the ancient Near East. Largely arid, it was a land of sheep, goats, and cattle. Nomads didn't keep pigs because they couldn't herd them through the desert. Villages in very dry areas didn't keep pigs because the animals needed a reliable source of water. Priests, rulers, and bureaucrats didn't eat pork because they had access to sheep and goats from the state-focused central distributing system and considered pigs filthy. Pigs remained important in only one place: nonelite areas of cities, where they ate waste and served as a subsistence food supply for people living on the margins.

This was the situation in the Near East around 1200 BC, when a tribe of people known as the Israelites settled in Canaan, west of the Jordan River in Palestine. Like most of their neighbors, the Israelites rejected pork. Unlike those neighbors, the Israelites came to consider pork avoidance a central element of their identity.

FOUR

"OF THEIR FLESH
SHALL YE NOT EAT"

God told Abraham to leave his homeland, Mesopotamia, and settle in Canaan. "I will indeed bless you," God told him, "and I will multiply your descendants as the stars of heaven." Those descendants, after a time, found themselves exiled in Egypt. There Abraham's heirs became so powerful that Pharaoh began to fear them and ordered all boys born to the Hebrews to be cast into the Nile. One mother set her infant son afloat in a basket of bulrushes, and he was rescued and adopted by Pharaoh's daughter, who named him Moses. When the boy had grown, God ordered him to lead the Hebrews back to Canaan. This Moses did, despite resistance from Pharaoh and a lengthy sojourn in the desert.

God then renewed the covenant with his people, issued the Ten Commandments, and gave highly detailed instructions on how the Hebrews must worship him. The greatest danger his followers faced was that they would make themselves impure and thereby force God to abandon them. He therefore laid out the rules of religious ritual and everyday behavior, including which animals were proper to eat. Among the forbidden beasts were pigs: "Of their flesh shall ye not eat, and their carcass shall ye not touch; they are unclean to you."

This is the story presented in Jewish scripture. Most of its details can't be independently confirmed, but archaeologists tell us that after 1200 BC, the human population exploded in the hills of central Israel and the Palestinian West Bank. These settlers were the Israelites—a mixed lot that, scholars speculate, included recent arrivals from Egypt, refugees from Palestinian cities, and seminomadic herders who chose to settle down.

Assume that the biblical story is true: Abraham was born in Mesopotamia, and the Israelites escaped from bondage in Egypt. The elites of both these great civilizations shunned pigs. They didn't sacrifice them to the gods, and they didn't eat pork. A formal ban on pork would have raised no eyebrows among the Israelites' neighbors. When the dietary laws of Leviticus and Deuteronomy were set down—most likely in the eighth century BC—very few people in the region ate pigs. Israelite priests, in banning pork, simply codified the beliefs of their place, their time, and their class.

Pork eating had become so rare in the Near East by this point that you might say that Jewish leaders banned something that didn't need banning. But ban it they did, and with far-reaching consequences. Because the ban was recorded as divine law, Jews maintained it after migrating to other regions of the world. And the Jewish ban influenced the pork prohibition within Islam, another Near Eastern religion that traces its heritage to Abraham. Today, there are about 14 million Jews in the world and 1.6 billion Muslims—meaning that the religions of nearly a quarter of the global population reject swine.

In the early years, the Israelite ban on pork wasn't all that important to the Jews: pork was rare in the region, so the issue rarely came up. Later, as pork-eating Greeks and Romans entered the picture, pigs took on a larger role. Avoiding pork, long a requirement of ritual purity, became a marker of cultural identity. Who were the Jews? They were the people who didn't eat pork.

There are many theories regarding why the Jews formally prohibited pork. Anthropologist Mary Douglas famously argued that Israelites deemed the cloven-footed, non-cud-chewing pig unclean because it was a "taxonomic anomaly." Like dirt, it was "matter out of place," threatening chaos by upsetting an orderly organizational scheme. Douglas's argument, though, suffers from circularity because the Israelites appear to have structured their taxonomic rules precisely to exclude the pig: perhaps the pig was unclean because it was a taxonomic anomaly, or perhaps it was a taxonomic anomaly because it was unclean.

Some have pointed to the unsuitability of pigs for desert conditions and the fact that pigs might devour foods, such as wheat and barley, that people needed for themselves. The pork prohibition therefore simply codified wise economic and environmental decisions. This view, although broadly correct, doesn't acknowledge that pigs can be kept on a small scale as urban scavengers, feeding off human garbage rather than challenging people for a limited food supply.

Others see in the pork ban the influence of pastoralist views, pointing out that some people who practiced long-distance herding held pig-keeping urbanites in contempt for their settled ways. This was the case with nomadic Mongols, for instance, who associated pigs with their sedentary Chinese enemies. But not all Israelites were nomads, and in any case there is little evidence of such views in the Near East.

Another theory holds that pigs were judged unclean because the Jews' idolatrous neighbors worshipped and sacrificed them. The evidence suggests, however, that others in Canaan sacrificed pigs only rarely. When they did, the sacrifices had links to the netherworld—that is, Canaanites sacrificed pigs not because they were clean but because they were dirty. And they certainly did not worship pigs.

One of the most persuasive recent theories points to the fact that the poor often raised pigs in order to gain control over their own food supply. A powerful central state, intent on controlling all aspects of the economy, would have seen such dietary autonomy as a threat to its control and a potential source of sedition. The elite banned pigs, in other words, so the poor would be hungry unless fed by the state. Pigs offered a dangerous independence and therefore had to be outlawed under cover of religious sanction. Though promising, this theory rests on a great deal of conjecture, and the best evidence for it comes not from ancient times, when the Israelite ban was codified, but from the rise of Islam in the seventh century AD.

Although the scholars who hold these theories tend to disagree on nearly everything, they come together on one point: all reject the view that the pork prohibition had anything to do with trichinosis. This theory found wide support starting in 1859, when scientists first proved the link between *Trichinella spiralis* and undercooked pork. It's not certain, however, that this parasite existed in ancient Palestine. And even if humans came down with the disease, they would have had a hard time connecting it with pork, because there's generally a ten-day delay between eating tainted meat and falling ill. Just about any kind of meat could make people sick—sheep can transmit anthrax, for instance—yet Jewish dietary law permitted other, equally dangerous types of flesh and singled out pork for prohibition.

Though the theory that the pork taboo was a public health measure has been thoroughly discredited, people have been reluctant to abandon it. Such beliefs stretch back at least to medieval times, and even then some authorities objected. "God forbid that I should believe that the reasons for forbidden foods are medicinal," wrote Jewish scribe Isaac Abrabanel. "For were it so, the Book of God's Law would be in the same class as any of the minor brief medical books."

This is a good reminder that there doesn't necessarily have to be a clear explanation for the pork ban. Scripture is, after all, primarily concerned with a people's relationship with their God, so to explain the Levitical dietary restrictions in terms of medicine or health or economics may miss the point. As Job learned after being stripped of his wealth and afflicted with boils, God wasn't much for explaining himself. Imposing an arbitrary food ban would not have been his most inscrutable act.

The pork prohibition, though, was far from arbitrary: it was thoroughly consistent with the sacred logic of the Bible and with God's command that the Israelites remain pure in order to preserve their relationship with him.

God demanded that the Israelites provide him with a physical home. They first constructed the Tabernacle, a portable structure housing the stone tablets inscribed with the Ten Commandments that the Israelites carried with them during their wandering in the desert. The Tabernacle was replaced by the more permanent Temple built by King Solomon in Jerusalem about 950 BC. The Temple occupied the central place in the religion of the Israelites. It preserved the presence of God among his people and ensured that blessings would continue to flow from him. To fulfill this role, however, the sanctuary had to remain pure, and it was under constant threat of defilement. If the people became polluted, they would pollute the sanctuary, and the relationship between God and his people would be severed.

To remain pure, the Israelites must eat only pure things. The eleventh chapter of Leviticus lays out the taxonomy: there were clean and unclean birds, clean and unclean insects, and clean and unclean fish. And then there were the land animals. The key rule was this: "Whatsoever parteth the hoof, and is clovenfooted, and cheweth the cud, among the beasts, that shall ye eat." Animals that "goeth upon paws"—dogs, wolves, lions—chewed no cud and divided no hooves and therefore were unclean. The same rule disqualified pigs: "The swine, though he divide the hoof, and be clovenfooted, yet he cheweth not the cud; he is unclean to you."

Diet played an important role in scripture even before God handed down the rules of Leviticus. Adam and Eve were vegetarians: in the book of Genesis, God gives them the grains, fruits, and vegetables to eat. Only after God had expelled them from the Garden and later scoured the earth with a flood did people begin to eat flesh. Meat, in other words, is the food of the fallen. God told Noah that he could eat "every moving thing that liveth" but then added a stipulation: "You shall not eat flesh with its life, that is, its blood." Noah's descendants were allowed to eat meat but never blood, which was thought to contain the "life force," or *nefesh* in Hebrew. Because the life or soul of an animal resided in its blood, to eat flesh with blood was to mingle life and death, two things that should be kept separate: "Eat not the blood: for the blood is the life; and thou mayest not eat the life with the flesh."

Just as God's people did not eat blood, they also did not eat animals that ate blood. Deuteronomy forbids eating carrion or "anything that dies of itself"—though it's lawful to "sell it to a foreigner" (caveat emptor)—presumably because the blood had coagulated within its veins and could not be drained. This explains why certain animals came to be declared unclean: they are predators and scavengers that eat the flesh of animals from which the blood has not been drained. Unlike Adam and Eve, the Israelites were no longer vegetarians—even so, they could eat only vegetarian beasts.

In specifying that God's people could eat only animals that chewed the cud and split the hoof, the priests displayed an intuitive sense of biological classification. Cud chewers, such as cows and sheep, are vegetarians. The pig didn't "chew the cud" because its gut had evolved to digest high-energy foods, including meat.

Scripture expresses disgust for the two most common scavengers, pigs and dogs. In the Christian Bible Jesus advises, "Give not that which is holy unto the dogs, neither cast ye your pearls before swine, lest they trample them under their feet, and turn again and rend you." "Dog" is an insult in the Bible, reserved for the most despised of people and linked to the animal's eating habits. "As a dog returneth to his vomit, so a fool returneth to his folly," we learn from Proverbs. According to the book of Kings, "Thus says the Lord: 'In the place where dogs licked the blood of Naboth shall dogs lick thy blood.'" In the next chapter: "So the king died . . . and the dogs licked up his blood." Such is the standard version of those verses. But in the Septuagint, a Greek translation of the Jewish Bible made in the second and third centuries BC, those dogs do not dine alone: "The pigs and dogs licked the blood of Naboth," and "the pigs and dogs licked up the blood" of the king. Some scholars speculate that this was the original version of the text.

Uncleanliness, in the Bible, is a contagion: predators and scavengers become unclean by eating bloody meat; men become unclean by eating the unclean flesh of animals that have eaten bloody meat; unclean men contaminate the temple so that God can no longer dwell with his people. That is why pigs, lickers of blood and eaters of carrion, could not be food for those who wished to remain pure: they were a vector for the unholy and would pollute anyone who consumed them.

Scriptural dietary rules grew more significant with time. When the laws of Leviticus and Deuteronomy were set down, few people in the Near East were eating pork. Archaeologists find no pig bones at all, or just a scattered few, in settlements from this period. Then, starting in about 300 BC, pig bones begin to appear in great profusion. The Greeks had arrived—and pigs would soon enjoy a renaissance after some nine hundred years of persecution.

Greek rule spelled major changes for the Israelites. The Greek king Alexander the Great had conquered the Persian Empire in 333 BC and taken over all the lands Persia had controlled, including Palestine. Whereas the Persians had worked through local rulers and allowed local peoples to live as they wished, the Greeks forcefully imposed Hellenistic culture on their subjects. In 167 BC the ruler Antiochus IV, a successor to Alexander, invaded Jerusalem and tried to stamp out Judaism, a story recorded in the Books of the Maccabees. The first book relates how Antiochus demanded "that all should be one people, and that each should give up his customs." Many Jews acquiesced and "sacrificed to idols and profaned the Sabbath." Worst of all, Antiochus ordered the Jews "to defile the sanctuary, . . . to sacrifice swine and unclean animals, and to leave their sons uncircumcised."

persecution
forced eating
pork

In the Second Book of the Maccabees, the invaders force pork into the mouth of Eleazer, an elderly Jewish scribe, but he spits it out. His tormenters, old friends who have gone over to the enemy's side, bring him aside and quietly tell him they will secretly replace the pork with kosher meat so that he can obey God's law while pretending to obey Antiochus. Again Eleazer refuses: "Many of the young should suppose that Eleazer in his ninetieth year has gone over to an alien religion," he says. "For the sake of living a brief moment longer, they should be led astray because of me." His purpose, he explains, is to leave "a noble example of how to die a good death willingly and nobly for the revered and holy laws." So he goes to the rack and is beaten to death over a mouthful of pork.

In the next chapter of the Second Book of the Maccabees, the pork-related punishments continue. A mother and her seven sons are arrested and told they must eat swine's flesh, but they too refuse. On the king's orders, a guard cuts out the tongue of one of the brothers, scalps him, and chops off his hands and feet. Then a large pan is heated over a fire, and the king orders his guards to take the brother, "still breathing, and to fry him in the pan," which they do. After he is dead, they kill another brother in the same way, and then another, until all seven brothers are dead, at which point Antiochus orders the mother slain as well.

Although these episodes occurred hundreds of years after the laws of Leviticus were laid down, they comprise only the second recorded instance of pork eating among the Jews. The first occurs in the book of Isaiah, when God expresses his fury at a few people who have eaten "swine's flesh, and broth of abominable things." They have done so in secret, hidden away in gardens and graveyards, and their sin is known only to God. It is a matter between the Lord and his people, and God promises to destroy the offenders.

Greek rulers killed the Jewish elder Eleazer because he refused to eat pork. For centuries, the scriptural prohibition played little role in Jewish life because all of the Israelites' neighbors rejected pork as well. Only with the arrival of pig-loving Greeks and Romans did pork abstention become a crucial aspect of Jewish identity.

In Maccabees, the situation is public. Infuriated by the Jews' desire to remain a separate people, Antiochus has outlawed the most visible symbol of their difference: their refusal to share a table with their neighbors. Here eating pork is not simply a matter of ritual purity, of remaining holy in order to keep the temple pure. It has become, instead, the key to cultural identity. The Books of the Maccabees provided a model of what it meant to be Jewish: even in the face of death, a Jew must refuse pork in order to remain true to his people.

Pork eating hadn't carried much significance as a marker of Jewish identity before the Greek conquest of Persia because most others in the region didn't eat pork either. Since the Israelites' return from exile in Egypt, abstaining from pork simply had been one way that they remained pure in order to preserve their relationship with God. Now, however, it also became a way that they drew boundaries between themselves and those they lived among. Indeed, when pork-eating Greeks ruled over the Jews, refusing pork became a key element of what it meant to be Jewish. You are what you eat, the saying goes, but the Jews were what they didn't eat.

The Jews rebelled against Antiochus and in 142 BC won control of Palestine and reconsecrated the Temple, an event commemorated in the celebration of Chanukah. Their independence lasted less than a century: in 63 BC the Romans conquered Jerusalem, and the Jews once more fell under the rule of pork eaters. Unlike the Greeks, the Romans responded to Jewish pork avoidance not with violence but with puzzlement and feeble jokes. Juvenal, the Roman satirist of the first century AD, noted that in Palestine "a long-established clemency suffers pigs to attain old age" because Jews "do not differentiate between human and pigs' flesh." It was said that Caesar Augustus, after hearing that King Herod of Judea had executed one of his own children, joked that he would "rather be Herod's pig than Herod's son."

There was a reason Jewish dining habits attracted attention: Romans loved pork with a passion matched by few people before or since. They developed the most sophisticated farming and breeding techniques that the world had ever seen and created elaborate—occasionally obscene—recipes to prepare pork for their lavish feasts. Such ostentatious pork consumption would only reinforce the divisions between Jews and Romans, and it would eventually establish pork as the meat of choice in the religion the Romans would help disseminate throughout Europe: Christianity.

FIVE

"MONSTROSITIES OF LUXURY"

An enormous pig, belly up, is wheeled into a banquet room in one scene of Federico Fellini's *Satyricon*. Trimalchio, the host, accuses the cook of roasting the animal without first gutting it and orders him whipped as punishment. The guests call for mercy, so Trimalchio demands, "Gut it here, now," whereupon the cook swings an enormous sword and slashes the pig's belly. The guests recoil in horror, but the steaming mass that pours forth is not the pig's viscera but cooked meat. "Thrushes, fatted hens, bird gizzards!" one character calls out. "Sausage ropes, tender plucked doves, snails, livers, ham, offal!" The dispute with the cook has been all in fun. The guests applaud, then grab hunks of meat and begin to gorge themselves.

Fellini's film, released in 1969, stays true to its source material, a work by Petronius written not long after the death of

Christ. In depicting Roman dining, Petronius satirized but did not exaggerate: there was no need to embellish the extravagant reality. The dish portrayed in the film, a medley of meats hidden within a whole hog, was known as *porcus Troianus*, or "Trojan pig," a nod to another great act of concealment. Petronius also describes a whole roast pig served with hunks of meat carved into the shape of piglets and placed along its belly, "as if at suck, to show it was a sow we had before us." Another feast featured what appeared to be a goose and a variety of fish, all carved from pork. "I declare my cook made it every bit out of a pig," the host exclaims. "Give the word, he'll make you a fish of the paunch, a wood-pigeon of the lard, a turtle-dove of the forehand, and a hen of the hind leg!" Why he should do so is left unexplained.

In cuisine, culture, and mythology, Romans delighted in concealment and disguise, metamorphosis and transformation, and in this they could hardly have been more different from the Jews. The Roman Empire formed a vast, cosmopolitan civilization that embraced and absorbed dozens of cultures. Few identities—whether of meats or of people—remained fixed. Trimalchio, in *Satyricon*, is a former slave who has won his freedom and then attained great wealth. A man calling himself a Roman citizen might have been born in northern Europe, Africa, or Asia Minor. Jews, by contrast, were dedicated to marrying among themselves, defending their small homeland, and preserving their ancient ways.

The differences between Romans and Jews extended to food. One people defined itself by rejecting pork, the other by embracing it. One called the pig abominable, the other miraculous. One saw the pig as a carrier of pollution, the other as a sign of abundance. Between them, Jews and Romans set the terms that would define the pig throughout the history of the West.

Pigs were the most common sacrificial animal in both Greece and Rome. They didn't pollute—they purified. In Greek mythology, after Jason and Medea kill Medea's brother, the enchantress Circe captures a piglet from "a sow whose dugs yet swelled from the fruit of the womb," slits its neck, and sprinkles its blood over the hands of the killers to remove the stain of murder. Similarly, a painted vase shows Apollo holding a sacrificed piglet, still dripping blood, over the head of Orestes, who has killed his mother. Priests killed a suckling pig to honor the gods before every public gathering in Athens. Romans killed pigs to seal public agreements, such as contracts and treaties, and to mark important private occasions, such as births and weddings.

Although the pig served as an all-purpose sacrificial animal, it carried a more specific meaning as a symbol of fertility. Demeter, Greek goddess of wheat, was honored with pig sacrifices. With her daughter Persephone—who was condemned to spend a third of each year in Hades—Demeter symbolized the circle of life, of death in winter followed by rebirth in spring. At Thesmophoria, the most widespread festival in ancient Greece, priestesses cast piglets into a pit and later retrieved their rotting carcasses and placed them on the altar of Demeter. The rotted pork was then scattered in the fields to ensure a good harvest. In Greece young pigs were known by the terms *khoiros* and *delphax*, both of which also could refer to women's genitalia, and the Latin *porcus* carried the same dual meaning. Aristophanes makes some horrifying puns on this double meaning in his play *Acharnians*, where a starving man disguises his two daughters as pigs and sells them in the market. The scholar Varro noted that Romans "call that part which in girls is the mark of their sex *porcus*" to indicate that they were "mature enough for marriage."

A swine, a sheep, and a bull are led to their deaths on a Roman altar. Whereas Jews rejected pigs as unclean, Romans sacrificed them to the gods and feasted on them with abandon. These two attitudes—Jewish repulsion and Roman embrace—have defined Western attitudes toward pigs ever since.

The use of pigs as fertility symbols traces back to the region's first farming communities. Just north of Greece in the Balkans, archaeologists have found early Neolithic statues of pigs studded with grains of wheat and barley. Like a seed germinating in the soil, a sow giving birth to many piglets demonstrated the bounty of nature. Sacrificing pigs honored the gods and ensured that the fields, and the people themselves, would enjoy abundant fertility.

Most people in the ancient world ate vegetarian diets heavy on grains and beans. This was the cheapest way to feed large populations. Rome was different. Although meat was expensive, Rome was rich, and a sizable class of people had enough money to eat it regularly.

Romans ate beef, lamb, and goat, but they preferred pork. Hippocrates, the Greek physician, proclaimed pork the best of all meats, and his Roman successors agreed. There were more Latin words for pork than for any other meat, and the trade became highly specialized: there were distinct terms for sellers of live pigs (*suarii*), fresh pork (*porcinarius*), dried pork (*confectorarius*), and ham (*pernarius*). According to the Edict of Diocletian, issued in 301 AD, sow's udder, sow's womb, and liver of fig-fattened swine commanded the highest prices of any meat, costing twice as much as lamb. Beef sausages sold for just half the price of pork. After the Punic Wars, the percentage of pig bones in Carthage doubled, just as it had in Jerusalem under Roman occupation: Romans kept eating pork even in arid climates such as North Africa and Palestine, where pigs were more difficult to raise.

The richest source on Roman cuisine, a recipe book known as *De re coquinaria,* or *On Cooking,* confirms this love of swine. Pork dishes far outnumber those made with other meats. The section called "Quadrupeds" contains four recipes for beef and veal, eleven for lamb, and seventeen for suckling pig. Other sections of the book offer recipes for adult sows and boars and nearly all of their parts, including brain, skin, womb, udder, liver, stomach, kidneys, and lungs. Archeology confirms that Romans carved up pigs more carefully and thoroughly than they did other creatures: pig skulls found in Roman dumps contain far more butchery scars than the skulls of sheep and cows, evidence that butchers excised the tongues, cheeks, and brains of pigs but not those of other beasts.

More than half of the dishes in *On Cooking* are relatively modest—barley soup with onion and ham bone, for example—and within the means of much of the urban population, but others demanded greater resources. Apicius is credited with inventing the technique of overfeeding a sow with figs in order to enlarge the liver, much as geese were stuffed with grain to create foie gras. In Apicius's recipe, the fig-fattened pig liver is marinated in *liquamen*—a fermented fish sauce central to Roman cuisine—wrapped in caul fat, and grilled. The recipe for pig paunch starts with this salutary advice: "Carefully empty out a pig's stomach." The cook is then instructed to fill the stomach with a mixture of pork, "three brains that have had their sinews removed," raw eggs, pine nuts, peppercorns, anise, ginger, rue, and other seasonings. Finally, the stomach is tied at both ends—"leaving a little space so that it does not burst during cooking"—boiled, smoked, boiled some more, and then served.

Some of the more elaborate dishes in *On Cooking* fall under the heading *ofellae*, which literally means a morsel of food. In one recipe, a skin-on pork belly is scored on the meat side, marinated for days in a blend of *liquamen*, pepper, cumin, and other spices, and then roasted. The chunks of meat would then be pulled from the skin, sauced, and served, forming bite-sized pieces that a diner could eat by hand while reclining, the preferred posture for Roman feasts. Another of the luxury dishes involves boiling a ham, removing the skin, scoring the flesh, and coating it with honey, a preparation that would not be out of place at Christmas dinner today.

Romans had a taste for blended milk, blood, and flesh that could make even a Gentile shudder. The Roman poet Martial had this to say about a roasted udder of lactating sow: "You would hardly imagine you were eating cooked sows' teats, so abundantly do they flow and swell with living milk." (Elsewhere, after a meal, Martial suffers the glutton's regret and remarks upon "the unsightly skin of an excavated sow's udder.") This preference veered into the bizarrely cruel. Some cooks, Plutarch claimed, stomped and kicked the udders of live pregnant sows and thereby "blended together blood and milk and gore," which was said to make the dish all the more delicious. The womb of this poor sow was eaten as well, with the dish called *vulva eiectitia*, or "miscarried womb."

Seneca, the Stoic philosopher and statesman, decried such dishes as "monstrosities of luxury," and he was far from the only critic. Roman rulers passed sumptuary laws limiting the amount that could be spent on meals and forbidding the consumption of items including testicles and cheeks. But the wealthy flouted such rules because the social hierarchy couldn't function without feasts: feasting provided the only way to learn who had grown richer and who had lost money, who was in the emperor's favor and who had been cast out. To curtail extravagance was to deny the very reason to feast.

Eating well had become central to the Roman self-image, and not just among the elite. Meat was so important that the empire got into the business of supplying it to a broad swath of the population free of charge. The emperor Augustus had started the practice of distributing free grain and bread as a way to ensure that the citizens remained well fed and peaceful. In 270 AD the emperor Aurelian started handing out free pork to those citizens already receiving free bread. By 450 AD about 140,000 citizens—a quarter or so of the city's population— were receiving the pork dole, five pounds a month for five months a year.

Free meat kept citizens happy but put heavy pressure on the empire's food supply: Rome's butchers had to process and distribute 20,000 pounds of pork every day. The emperor ensured an adequate supply by imposing a tax, payable in pigs, upon certain forested areas of Italy south of Rome. In some years these regions sent more than 30,000 pigs to Rome. Producing and distributing that much pork was no simple matter, but the empire's farmers managed it with relative ease.

To feed the empire's expanding population, Rome created the most sophisticated agricultural system the world had ever known. Previously, farming had been a local affair. Even in the great civilizations of Egypt and Mesopotamia, production and consumption occurred within a fairly circumscribed area defined by irrigated river valleys and surrounding rangeland. By contrast, imports from outside the Italian Peninsula constituted three-quarters of Rome's food supply.

Rome brought all of the Mediterranean world and much of Europe within its orbit, pulling in grain from Egypt, cured meats from Spain, olive oil from Syria, and spices from further east. The wheat that satisfied Caesar's bread dole was mostly imported from North Africa, where it was collected as tax. Grain sufficient to feed hundreds of thousands of people moved around the region by ship and filled large granaries that provided insurance against famine.

Although Romans imported grain by ship, they raised nearly all of their livestock within Italy. They kept sheep primarily for wool and secondarily for milk and cheese. Goats were rare, though sometimes raised for milk. Cows offered dairy products, and oxen pulled plows in the fields and carts on the road. Meat from these animals was eaten, but it was usually a by-product rather than the principal reason for raising them. Archaeologists tell us that most butchered cattle show stress injuries to their leg bones, meaning that they worked hard before ending up in the pot. Beef and mutton came from older animals—ewes and cows whose udders had dried up, rams and bulls who had become infertile, and oxen that could no longer pull a plow.

Only pigs were raised exclusively for food. They were eaten when young and therefore were far more tender than worn-out oxen. A popular saying held, "Life was given them just like salt,

to preserve the flesh"—meaning that pigs had no reason for living other than to feed people. Given how much Romans loved
to feast, this was no small consideration. According to Varro,
Rome's most important agricultural writer, "the race of pigs is
expressly given by nature to set forth a banquet."

Romans created the first detailed farming manuals, which
devoted special attention to pigs. The authors likely adapted
farming techniques from Greece, but Greek writers were too
refined to dirty their hands with practical advice. The best
glimpses into Greek farms therefore appear in literature: *The
Odyssey* describes Ulysses's swine farm, a large operation that
involved fifty sows tended by the beloved swineherd Eumaeus.
Romans, by contrast, were generous with explicit husbandry
advice. Varro devoted more attention to pigs than to cows,
sheep, or goats and suggested that to do without pigs was unthinkable: "Who of our people cultivate a farm without keeping
swine?" Pigs were the perfect meat-producing animal. Because
they were raised only for food, they could be bred for flavor
and weight-gaining ability rather than strength or milk production. Columella, writing in the first century AD, extolled
their versatility: "Pigs can make shift in any sort of country,"
finding suitable pasture "in the mountains and in the plains."
The writers offered feeding advice for each stage of a pig's life,
from piglet to lactating sow.

Roman writers paid special attention to breeding. Boars,
Columella tells us, should possess "huge haunches" and be "as
lustful as possible when they have sexual intercourse." He specified similar qualities for sows and described how to build their
sties and provide clean bedding for comfort. They were bred
twice a year, gestating for just under four months, nursing for
two, then starting over again. Some farmers kept herds of three
hundred or more sows, which meant they produced thousands

of pigs for market every year. With that sort of production, farmers had the incentive—and the means—to breed the perfect pig.

Or, as it turned out, two perfect pigs. Bones from Roman
dumps indicate that most pigs stood sixty to seventy centimeters
at the withers. Another group, smaller in number but larger in
stature, stood about eighty centimeters. The shorter pigs were
scattered all across the Italian Peninsula, while the taller type
clustered around Rome. Roman writers confirm the existence of
the two types. The smaller looked like a downsized wild boar:
rangy, long-legged animals with what Columella called "very
hard, dense black bristles." This type lived in the regions south
of Rome that produced pork for the public dole, wandering
the forests to eat acorns, nuts, and other wild foods. They liked
to "root about in the marsh and turn up worms," Columella
wrote, and "tear up the sweet-flavored rootlets of underwater
growths." The best feeding grounds for such pigs, he advised,
were forests with "cork oaks, beeches, Turkey oaks, holm oaks,
wild olive trees," as well as plum and other fruit trees, for such
trees "ripen at different times and provide plenty of food for the
herd almost all the year round."

Columella also described the larger variety, "smooth pigs
and even white ones." This latter type lived in sties, so it didn't
need agile legs for running through the woods or thick bristles
to keep warm. Farmers fattened these pigs on wheat, barley,
lentils, and chickpeas. Varro reports that nursing sows were fed
"two pounds of barley soaked in water" daily. These fat white
pigs were kept closer to Rome to feed the city's gourmands: a
feast in *Satyricon* features "three white hogs." The white sows
also birthed the suckling pigs that Roman diners prized. Columella advises that on all farms "near towns, the suckling pig
must be turned into money" by selling it to elite households.
Adult white pigs were sacrificed to the gods; sty-fed and slow

of foot, they were more docile than forest pigs when facing the priestly axe. And they made impressive offerings, not only because they were expensive but also because they were white, the preferred color of the gods.

The gods demanded white because white suggested purity, and the way Roman pigs lived helps explain how they could be considered pure. In the Near East many pigs lived as scavengers on city streets, devouring garbage and human waste, and earned a reputation for filthiness. Rome was a cleaner place: aqueducts brought clean water, and sewers carried away filth. The Italian Peninsula, moreover, enjoyed enough rainfall to create marshes and oak forests, and trade networks brought an abundance of wheat and barley. Rather than eating carrion and garbage, Roman pigs spent their days devouring nuts in the woods or grains in the sty.

Diet had a profound effect on the pig's flavor. In the guts of cows, sheep, and other ruminants, microorganisms digest and transform fatty acids, so what the animal eats has less influence on its flesh. Pigs, not being ruminants, lack those microorganisms, so they deposit fat in the same form they ingest it. A pig that sups on fish guts will taste very different from one that eats hazelnuts. Compared to their scavenging cousins in the Near East, Roman pigs ate well, stayed clean, and tasted delicious.

It took time, energy, and wealth to create such flavorful pork—and the sophisticated Roman system of production was able to expend all three. Rome became the first large city where tens of thousands of people had regular access to meat, and this did not come cheap. Roman pigs competed with humans for food: every pound of barley fed to a pig was a pound that didn't feed a person. Fattening livestock on grain is an inefficient way to produce calories, and the practice was quite rare globally before about 1800. Rome was the exception. Because it controlled the region's food supplies, the empire could afford to feed both its pigs and its people. A sophisticated economy created vast wealth, and that wealth allowed Roman pigs to grow fat.

Lean times lay ahead. When the Roman Empire fell, its white sty pigs fell with it. Rome's small, bristly woods pigs, by contrast, landed on all four feet. They were perfectly adapted to the rough conditions of the dark medieval forest, where they would earn the respect of new generations of farmers, cooks, and diners.

SIX

THE FOREST PIG

In 401 AD an army of Goths swept from the Balkans into north-ern Italy. Soon other Germanic tribes forded the Rhine River and invaded Gaul, a region first conquered by Julius Caesar 450 years earlier. The invaders—Romans called them "barbarians"—roamed freely through the empire, capturing more territory until they finally deposed Emperor Romulus Augustulus in 476 AD. The Gothic tribes then divvied up western Europe: Anglo-Saxons in England, Visigoths on the Iberian Peninsula, Vandals in North Africa, Ostrogoths in Italy, Franks and Burgundians in France, Alamans in Germany. The Eastern Roman Empire—ruled out of Constantinople and often called the Byzantine Empire—survived for another millennium, but the Western Roman Empire was dead.

The fall of Rome, in the traditional view, plunged Europe into the Dark Ages—a period devoid of art, literature, fine din-ing, clean water, and other luxuries—from which it emerged only with the first glimmers of the Renaissance 1,000 years later. Historians more recently have proposed that there was no sudden fall from civilization to barbarity but rather a gradual transition in which the Roman and Germanic worlds blended to create new, not necessarily inferior, cultures. The jury is still out on that larger argument, but this much is true: the fall of the Roman Empire brought rapid change to the world of pigs.

Only one type of Roman pig survived the collapse. Rome's complex networks of Mediterranean commerce disintegrated alongside the empire itself. With the disappearance of that trade and of the concentrations of wealth it had produced, there was little market for suckling pigs or large white swine. It's hard not to see such pigs as symbolic of Rome as a whole, grown fat and lazy on the spoils of empire. Archaeologists digging in post-Roman sites don't find any bones of large swine. Only the rangy black pigs survived. In the chaos of the empire's fall, they snuck off into the woods to shift for themselves and soon reemerged at the heart of European culture—as the staple source of meat and fat for both rich and poor.

Pigs had long been at home in northern Europe. The region enjoys the benefits of the North Atlantic Drift, a powerful ocean current that brings ashore warm winds and year-round rains that encourages the growth of hardwood trees. Before agriculture and metal axes reached Europe, Paleolithic tribes huddled along riverbanks and seacoasts because the rest of the landscape was thick with forbidding forests, home to wolves, bears, and the Eurasian wild boar, *Sus scrofa*.

By about 7500 BC, those wild creatures had come to share the northern European woods with domestic animals imported from the Near East. The populations of that area's first farming communities had grown quickly, and only migration could relieve the pressure. One group of Near Eastern farmers, traveling by boat, hopscotched along the northern coast of the Mediterranean, rounding the Iberian Peninsula in the eighth millennium BC. Another group moved overland out of Turkey and Greece, following the valleys of the Danube and Elbe Rivers and settling in central Europe at about the same time. These farmers, who soon wiped out or absorbed the local hunter-gatherer tribes, brought with them the full range of Near Eastern livestock: the bones of sheep, goats, cows, and pigs have been found at the earliest sites where they settled in Europe. Not all of these animals, however, survived in their new habitat.

Genetic studies tell us that the first wave of imported Near Eastern pigs died out and was replaced by a new strain domesticated from the wild boars native to European forests. This domestication event likely mirrored those that had happened earlier in China, the Near East, and elsewhere, when wild boars crept out of the woods to scavenge in human settlements. With guidance from Europe's farmers, who had prior experience tending livestock, some of these wild creatures evolved into an entirely new—and yet not new at all—variety of animal: *Sus scrofa domesticus*, almost precisely like their cousins in the Near East but descended from a different stock of wild boars.

Livestock, pigs included, sifted themselves by climate and terrain in this changing European landscape. Goats and sheep predominated in highland regions and in dry Mediterranean lands. Cows grazed on the thick grasses of Europe's northern fringe. Pigs reigned wherever forests remained intact. The pig-based fertility religions of the ancient Mediterranean—the same ones that gave rise to the cult of Demeter—traveled north with the first farmers. Among the Celts who occupied much of Europe, swine became symbols of war, fertility, and feasting. Celtic warriors adorned their helmets with boar bristles, and in *Beo-wulf* the hero wears a golden helmet ornamented with images of boars. In Norse mythology the fertility god Frey sports a mighty phallus and rides a golden-bristled boar, and at the festival known as Yule—later merged with Christmas—worshippers sacrificed a boar to Frey to ensure a good harvest. An Irish myth tells of pigs that were slaughtered and devoured and then, a day later, sprang back to life to be killed and eaten again—a fantastical exaggeration of the genuine fecundity of pigs.

The historical record contains a few traces of the pig-keeping practices of early Europe. In northeastern Gaul—parts of the Netherlands and Belgium today—a people known as the Salian Franks came to power and established a legal code just after the fall of Rome. They had two laws for goats, five for sheep, fourteen for cattle—and twenty for pigs. The code specified the fine for stealing more than fifty pigs, indicating that swine rustling was no minor problem. It imposed a higher fine for stealing a pig from a sty than from a field and a higher fine still if the swineherd was present when the theft took place. Other laws addressed "he who steals a leader sow" (presumably one that led other pigs into the forest to forage), "he who steals a bell from another man's troop of pigs," and "he who steals a sacrificial gelded boar [that] had been consecrated." From these laws we can infer that the Salian Franks sacrificed boars to their gods, kept pigs in sties to protect them from thieves and predators, and herded them through fields and forests.

The Anglo-Saxons, the Germanic tribe that had swept into England in the fifth century, wrote laws to protect both pigs and their forest grazing lands. Anglo-Saxons valued a pig at twice the price of a sheep and fixed severe penalties for destroying acorn-producing trees. Mast—the fruit of oak, beech, chestnut, and other trees—was the most valuable forest product. In a practice known as *denbera* in Saxon and *pannage* in Norman, the nobles who controlled the forests charged for the right to fatten swine in the woods each fall. Throughout Europe the size of a forest sometimes was judged not by its acreage but by the number of pigs it could support. In England's Domesday Book (1086 AD), a sort of census of the kingdom, designations such as "wood for 100 swine" served as measurements for some forests. In ninth-century Italy a monastery's forest was judged to be 2,000 pigs big. Whether the forest was five or fifty square acres mattered less than the number of swine it could feed, because that determined its worth.

Some pigs spent their entire life cycle in the woods. The tips of stone arrowheads have been found embedded in the bones of domestic swine from Neolithic England, suggesting the animals were kept in a semiferal state and hunted down when needed. In Europe "hogs run wild," wrote the Greek historian Strabo, whose *Geography* describes his travels in the time of Augustus. These free-ranging domestic animals could be every bit as fearsome as their wild brethren, which in their various habitats were known to fight off large predators like tigers, crocodiles, and bears. "It is dangerous for one unfamiliar with their ways to approach them," noted Strabo, "and likewise, also, for a wolf."

Some forest pigs were closely managed. In the forests of Kent in the ninth century, pigs lived for most of the year on the manors, then were driven in the autumn to seasonal settlements known as *denns*—the tradition survives in place-names such as Tenterden—where they grazed on mast. The swineherd contracted with farmers and gathered up five or six hundred pigs, for which he was paid by the head. Assisted by a herding dog, he drove them to the forest, where he built a rough pen under a large tree and filled it with straw and ferns for bedding. Then he would feed the pigs, blowing a horn while they ate so they would associate the sound of the horn with food. He would turn them out to forage during the day, then call them back to the pen for the night with a blast of the horn. Swineherds carried either a long, slender pole for smacking branches to bring down acorns or a short, stout stick, flung up into trees for the same purpose.

A Roman farmer who raised suckling pigs for banquets might have looked with horror upon such methods, but the forest pig was perfectly adapted to the conditions of the Middle Ages. The European forest was no place for coddled sty pigs. Husbandry here was defined in roughly equal measures by human intervention and natural selection. Pigs competed with each other for the nuts that dropped to the ground, and wild boars still roamed the woods, muddling the gene pool by interbreeding with their domestic cousins.

In their embrace of swine, medieval Europeans had much in common with ancient Romans. Nobles saw Rome as the pinnacle of civilization and sought to establish a similar heavenly empire on earth. They hunted boar as the ancient Romans did and feasted on pork just as ravenously.

Above all else, the era's warrior culture valued courage and bravery, which noblemen could demonstrate on the battlefield and in the hunt. The most prized quarries were boar and deer, and only nobles were allowed to kill them. An English law of 1184 decreed that commoners who poached these animals would be punished with blinding and castration. European hunters viewed boar and deer as polar opposites. The deer was elegant and swift, a test of the hunter's speed and cleverness.

Swineherds depicted in a fourteenth-century English manuscript knock down acorns for their pigs, bristle-backed animals that roamed the forests and sometimes interbred with their wild-boar cousins. Medieval Europeans rivaled the Romans in their love of swine. (Courtesy British Library)

The boar, powerful and ugly, was impervious to pain and fought fiercely at bay, demanding strength and bravery from the hunter. Gaston Phoebus, in his fourteenth-century treatise on hunting, called the boar the fiercest of all animals. Lions and leopards kill with claws and teeth, while "a boar kills with a single stroke, as one might with a knife."

Classical Greece and Rome had shaped those views through legend and myth, such as the tale of the Calydonian Boar. The king of Calydon, the story goes, made offerings to the gods but neglected Diana, who expresses her fury by sending a wild boar to ravage his kingdom. When the greatest warriors of Greece gather to hunt him, "the boar rushes violently into the midst of the enemy, like lightning darted from the bursting clouds," Ovid writes in *Metamorphoses*. The boar slashes at an approaching hero, and the man's "bowels, twisted, rush forth, falling with plenteous blood."

King Arthur too hunted a mythical boar. In a Welsh tale from early Anglo-Saxon times, Arthur and his fellow warriors tracked the boar across the Irish Sea and then engaged him in a nine-day battle that "laid waste to the fifth part of Ireland." After the fight the boar swam back across the sea, shook the saltwater from his bristles in Wales, and again began killing men by the dozen. Finally, in Cornwall, Arthur cornered the boar and drove him into the sea, never to be seen again. Arthur became known as the Boar of Cornwall for his bravery.

As Arthur's epithet suggests, killing a wild boar came to be considered a mystical act that transferred the strength of the animal to the hunter. Domestic pigs—which, in their tusked, shaggy, semiferal state, looked much like the wild boars with whom they shared the woods—basked in the reflected glory of their wild cousins. In present-day Belgium, bones dug up at castles and monasteries show that nobles and monks consumed a lot of pigs, while peasants, when they could afford meat, ate mostly cattle and sheep. Archaeology in England shows the same pattern: commoners ate beef and mutton from older animals culled after the end of their productive lives. The trash heaps of the elite—in castles, palaces, monasteries, and convents—were piled high with pig bones.

Medieval cooks also mimicked Roman styles of preparation. Many recipes were derived from Apicius, whose manuscripts were copied in medieval monasteries and courts. Meat was boiled or spit-roasted and served with heavily spiced sauces. Because the spice trade with the East had expanded enormously by the high Middle Ages, European noblemen had far more potent spices at their disposal—such as cinnamon, cloves, and nutmeg—than did the Roman elite. (Medieval Europeans ate spices because they liked them, not to mask the taste of bad meat: an entire pig could be bought for the price of a pound of pepper, and anyone who could afford spices could also afford fresh meat.) Medieval cooks also borrowed from Rome the habit of cobbling together strange creations: they would cut both a suckling pig and a capon—a castrated cock—in half, then sew the forequarters of one onto the hindquarters of the other. The Count of Savoy's chef presented a boar's head set between its disarticulated feet, with one side of the head covered in gold foil and the other glazed with green sauce, like a heraldic symbol. A camphor-soaked wick was placed in the boar's mouth and lighted, so the boar was served breathing fire. One cookbook offered a recipe for a roasted rooster, wearing a tiny helmet and carrying a lance to match, sitting astride an orange-glazed suckling pig.

The pig also played humbler roles in medieval kitchens. In noble houses, the pantry of preserved foods became known as the "larder" because lard, which at the time referred to rendered fat or any fatty cured pork, was the most important item it held. This was another miracle of pigs: they were not only suitable for feasting but also, when preserved, provided a store of food for lean times.

Curing, at its most basic, involves nothing more than drying meat. Bacteria requires moisture to grow, so the drier the meat, the less likely it is to rot. In arid climates meat can be cut into strips and left to cure in the air; Norwegians preserved cod this way, and Native Americans did the same with venison. Usually, though, curing involved salt. Coating a piece of meat with salt creates osmotic pressure: water rushes out of the animal cells toward the salt, drying out the meat. Salt is also directly toxic to bacteria, killing them through osmosis by sucking the moisture out of them. Sometimes the salt gets an assist from wood smoke, which deposits a variety of bactericidal compounds on the meat's surface, along with delicious flavors. Any meat can be cured with salt, but lean meats like beef tend to become tough when so preserved. Cured pork, with its generous veins of fat, remains tender.

The ancients understood the practice of curing, if not the science behind it. Greeks used the same word to describe both the curing of pork and the Egyptian practice of mummification, because drying out a dead pharaoh was not so different from preserving a leg of pork. Roman farming manuals record the earliest detailed instructions for treating the latter: pour a layer of salt into the bottom of a large pottery jar, place hams, skin side down, on top of the salt, and cover the meat with more salt. Then add alternating layers of hams and salt until the jar is full. After five days, remove the hams and repack them, with the top layer of hams now on the bottom. After twelve more days, remove the hams, brush off the salt, dry "in the breeze" for two days, rub down with oil and vinegar, cold-smoke for two days, and then hang in a meat house. According to Cato, "No moths nor worms will touch" hams prepared in this way.

During the height of the Roman Empire, some of the most highly prized cured pork on Roman tables was imported from the European provinces. Varro insisted that the Gauls of southern France made the best bacon. Cato reported that a Gallic group from northern Italy cured 3,000 or 4,000 hams annually for export to Rome. These Gauls lived around Parma, now famous for its prosciutto, which suggests that the region has enjoyed a continuous tradition of ham making for two millennia. The same is true for Iberia and Germany. Varro recommended pork from what is now Portugal, and Strabo reports excellent hams from the Spanish side of the Pyrenees. Martial gave a nod to hams made along the Rhine, in the same region where Westphalia hams are now made.

For the millions of farmers who cured pork for their own use, lard was as important as meat. Living things need to eat fats, which help create the membranes of all cell walls and provide an efficient food source, packing twice as many calories as an equivalent weight of sugars or starches. Ancient cooks often boiled their meat because this method, unlike spit-roasting, preserved the fat for later use. Fat was so rare and precious that an old Hittite law code specifies that if a dog eats lard from a man's kitchen, he can legally kill the dog and rescue the lard from its stomach—and then, presumably, eat the lard.

The Mediterranean world harvested most of its dietary fat from olive trees. Since northern Europe was too chilly for olives, the pig functioned as a sort of olive tree on the hoof. A lard belt stretched across Europe, just north of the olive oil belt and overlapping with the butter belt along the cow-heavy coast of the North Sea. Pigs feasted on nuts in the fall, putting on a few inches of subcutaneous fat to live on through the winter. Humans intervened in this process, killing the animals and using the fat and meat as their own winter provisions. Medieval calendars, decorated with illustrations depicting the usual occupations for each month, devoted the fall to pigs. The October illustration typically showed a swineherd in the woods with his pigs, the oak branches heavy with acorns. November, known as "blood month" in Anglo-Saxon, depicted a pig slaughter. The feast day of Saint Martin, November 11, became an important holiday because it marked the start of the slaughter season, when the weather turned cold enough for meat to cure before it spoiled.

For medieval Europeans, the seasons of the year were a bumpy cycle of warmth and cold, abundance and scarcity, but pigs smoothed the ride: they were fattened on the fruits of summer and fall and then slaughtered for winter sustenance—while a pregnant sow, bedded down in a warm shed, promised a fresh crop of piglets in the spring. Many proverbs indicated that a supply of salt pork represented safety: "He who has barley bread, and fatback for his gullet, can say that he is happy."

Not everyone, however, felt quite so positively about the pig. As Europe's human population grew and trade expanded, the region's forests gradually disappeared, and the acorn-loving woods hog was forced to find other ways to sustain itself. Its new food source mirrored that of the very first domestic pigs: human waste, scavenged from the streets. In the cities of Europe, as in those of the ancient Near East, people found such habits troubling—and the pig's reputation suffered accordingly.

SEVEN

"SWINE EAT THINGS CLEAN AND UNCLEAN"

Sometime around 1210 AD, Francis of Assisi and his companion Friar Juniper paid a call on a sick friar and asked if he needed anything. The man told his visitors he was hungry for a pig's foot. Friar Juniper immediately snatched up a knife—"I believe 'twas a kitchen knife," his hagiographer tells us—ran toward a herd of pigs, and "falling on one of them, cuts off a foot and runs away with it." Friar Juniper then cooked the foot and fed it to the sick friar while telling the patient "with great glee" about "the assaults he had made on the pig."

The swine's owner, understandably, complained to Saint Francis, who upbraided Friar Juniper for theft. The friar then explained to the pig's owner that he had acted only out of concern for the sick man. The owner, moved by the friar's humility, forgave him and donated "what was left of the pig" to the

As Europe's forests were felled to grow crops, pigs took up residence in towns, as depicted in this detail from Breugel the Elder's 1559 drawing *Fair at Hoboken*. The scavenger pig's diet, which included the occasional human corpse, contributed to a decline in the reputation of pork in the late Middle Ages.

monastery. Saint Francis thus managed to resolve the dispute, but he did so by treating it as a property crime, no different from stealing a loaf of bread. The patron saint of animals expressed no sympathy for the pig hobbling about the woods on a bloody stump.

According to another account of Saint Francis's life, he once was staying at a monastery when a sow came across a newly born lamb and "slew him with her greedy jaws." The sow had done what pigs do—eating any tasty morsel that presents itself—but Francis judged beasts by the moral standards of the church. "Cursed be that evil beast," the saint said, and the sow died three days later.

These stories of Saint Francis reflect a broader shift in attitudes toward swine. Medieval nobles, hunting boar in the woods and feasting on domestic pigs, carried on the Roman tradition of swine love. But the Jewish distaste for pigs persisted as well, transformed but preserved within European Christianity.

Although Christians ate pork, many of them retained the Jewish prohibition against eating carnivores and scavengers. A pig could be eaten, but only if that pig had not dined on nasty things. As the centuries passed, clean-living pigs became harder to find. Europe's population grew rapidly in the twelfth and thirteenth centuries, and trees were cleared to plant crops. The pig, having lost its forest home, found a new one in the growing towns and cities, where its eating habits again came under scrutiny. To the old indictment—that pigs ate rotting animals and other filth—were added new charges that pigs devoured human corpses and killed children. Pigs were not only unclean: on occasion, they seemed downright evil.

The lamb stood meekly in the top spot of the Christian Bible's hierarchy of animals, and the swine wallowed at the bottom. According to the New Testament, Christ was "the Lamb of God, who takes away the sin of the world," as well as the Good Shepherd who cares for his flock. The Christian Bible picked up this theme from the Jewish scriptures. "The Lord is my shepherd, I shall not want," reads Psalm 23.

Pigs have less positive associations in the New Testament. According to the Second Epistle of Peter, those who turn their back on Jesus call to mind a proverb: "The dog turns back to its own vomit, and the sow is washed only to wallow in the mud." The prodigal son, after squandering his inheritance, must accept the most abject work imaginable: he feeds another man's pigs and grows so hungry that he wishes he could have "filled his belly with the husks that the swine did eat." (Those husks, incidentally, were likely pods of the carob tree and certainly not ears of corn, a New World crop unknown in Eurasia at the time.)

Jesus himself had little love for pigs. While traveling among the Gaderenes near the Sea of Galilee, he came across a man possessed. He said to the demons, "Go," and the demons went out of the man and entered a herd of pigs: "Behold, the whole herd of swine ran violently down a steep place into the sea, and perished in the waters." The swine numbered 2,000, and yet no one mourned their loss. Jesus would rescue one lost sheep, but he sent thousands of swine to their deaths.

We should not be surprised. The psalm does not say, "The Lord is my swineherd, I shall not want." The prodigal son's father, upon welcoming him home, does not kill the fatted hog. Jesus was born a Jew and died a Jew, and he passed along to his followers the Jewish view of swine.

These prejudices, common in early Christianity, were transmitted down through the centuries. European works known as bestiaries, which circulated widely in the twelfth and thirteenth centuries, drew Christian lessons from the behavior of animals. The lamb was entirely blameless. In the words of one bestiary, the lamb represents "our mystic Saviour, whose innocent death saved mankind," as well as any Christian "who obeys his mother the Church." All other animals have a mix of good and bad attributes, except for the pig, which is represented in entirely negative terms. "The pig is a filthy beast; it sucks up filth, wallows in mud, and smears itself with slime," one bestiary claimed. "Sows signify sinners, the unclean and heretics." In addition to filth, the pig stood for gluttony and lust, the last two of the seven deadly sins.

There is a glimmer of biological truth to some of these charges. Pigs have few sweat glands, so they wallow in mud and let evaporative cooling do the work of thermoregulation. They are often in a hurry to eat, but that is a by-product of their diet: whereas sheep eat foods that are abundant in nature—grass and leaves—pigs need energy-intensive foods that tend to be scarce, demanding quick action that might resemble gluttony.

Nature also made pigs lustful. During sex the boar's penis— two feet long, thin as a pencil, and corkscrew shaped at the tip— locks into a corresponding twist in the sow's cervix, and there the two remain, for fifteen minutes or longer, during which time the boar ejaculates up to a pint of sperm. An early agricultural writer described pigs as "very lecherous, and in that act tedious." Europe's pagan cultures—the Celts, Greeks, and pre-Christian Romans—had celebrated swine for their exuberant fertility; Christians had a more troubled relationship with sex.

To modern eyes it's difficult to blame the pig for doing what nature demands. But to medieval Christians, brought up to find moral lessons in the natural world, the pig's habits made it a problematic choice for the dinner table.

The New Testament freed Christians from most Jewish dietary laws, with Acts of the Apostles the key text. Peter, while "very hungry," experiences a vision of "animals and reptiles and birds of the air," and a voice tells him to "kill and eat." Peter protests that he has "never eaten anything that is common or unclean," but the voice tells him, "What God hath cleansed, that call not thou common."

Just as Jews defined themselves as not-Greek by refusing to eat pork, Christians defined themselves as not-Jewish by eating it. At one of the councils of Antioch, the church fathers recommended that Christians eat pork precisely because the "synagogue execrates" it. Eating pork became a symbol of the New Law. Rather than a small tribe content to remain pure in its own homeland, Christians sought to convert the entire world to the Gospel of Jesus, and they were happy to welcome lovers of swine. It is hard to imagine the Roman Empire embracing Christianity if Christianity had not first embraced pork.

And yet food rules proved hard to throw off entirely. Even the Acts of the Apostles hedged its bets, telling Christians they must "abstain from meats offered to idols, and from blood, and from things strangled." This preserved the Levitical law against mingling flesh and blood, the very concept that had led to the ban on pig flesh. The laws of Anglo-Saxon England specifically outlawed the consumption of strangled animals, and Christians generally embraced the Jewish prohibition against eating carnivores like bears, wolves, and cats—animals that ate flesh from which the blood had not been drained and were therefore polluted. An Irish text warned people not to eat "a scab from one's own body": since humans eat meat, this mild self-cannibalism constituted eating the flesh of a predator.

The pig's omnivorous diet, combined with the heavy symbolic freight the animal carried, prompted anxiety. Irish priests created the most elaborate pork regulations. If a swine ate carrion just "once or twice," it could be eaten once that carrion was "ejected from its intestines," according to seventh-century rules known as the Canons of Adamnan. That's assuming that the carrion in question was not a human corpse. "Swine that taste the flesh or blood of men are always forbidden," the rules warned.

Other Christians were more permissive toward pigs that had dined on people. Theodore of Tarsus, who became archbishop of Canterbury in 668 AD, set down another set of dietary rules. If swine merely tasted human blood, they remained clean, but if they "tear and eat the corpses of the dead, their flesh may not be eaten until they become feeble and weak and until a year has elapsed." By then, the fat derived from the human flesh would be purged, and the pig could be fattened on clean food and eaten.

Medieval swine had frequent opportunity for scavenging human remains. Medieval armies could be slow to collect their dead after battles, and executed prisoners, suicides, and people excommunicated from the church were often left unburied as a form of punishment. In Shakespeare's *Richard III*, the title character is described as a "foul swine" who "Swills your warm blood like wash, and makes his trough / In your embowell'd bosoms." Such habits of pigs necessitated caution by those who ate pork. "Cows feed only on grass and the leaves of trees," the Canons of Adamnan noted, but "swine eat things clean and unclean."

"Clean" food for pigs became harder to find as the medieval era progressed. In 1000 AD most of the great European forests where pigs foraged remained intact. Over the next two hundred years, Europe changed dramatically as farmers captured coastal marshes from the sea and, further inland, cut down trees to plant wheat, barley, and rye. The process was nudged along by the Medieval Warm Period, roughly from 950 to 1250 AD, when average temperatures rose a degree or two and growing seasons lengthened.

As towns and cities grew, pigs came in from the dwindling forests and took up residence in the streets. We know about this mostly from attempts to banish pigs from cities. In 1131 in Paris, a boar ran under the legs of a horse ridden by young Prince Philip, causing the boy to be thrown to the street and killed. In response, Parisian authorities banned pigs from running loose in the city. Similarly, in 1301 the English city of York passed an ordinance reading, "No one shall keep pigs which go in the streets by day or night, nor shall any prostitute stay in the city," thus drawing an equivalence of sorts between unrestrained animals and unrestrained women.

The frequency with which cities issued and reissued prohibitions on pigs indicates that the bans didn't work, and they
didn't work because pigs played a necessary role. In Paris and
most other towns and cities in Europe, the rule governing sanitation was *tout à la rue*, "everything in the street," which meant
Parisians simply flung garbage and the contents of their cham-
ber pots out the window. The wealthier might have pit latrines
or cesspools, but the men who cleaned them often dumped the
waste in the gutter. Sometimes this filth was collected as fertilizer or as a raw material for making gunpowder, but much
of it found its way into the stomachs of pigs. A set of German
playing cards from 1535 depicted pigs roasting excrement on
a spit and then eating it. The theologian Honorius of Autun,
writing in the eleventh century, describes wicked people as "shit
for the stomach of pigs," a metaphor that reflects the animal's
dining habits. In an English text, a woman explains that she
won't serve pork because the local pigs "eat human shit in the
streets."

And that wasn't the pig's worst offense.

The court records of medieval Europe record dozens of cases
in which pigs were tried and convicted of attacking children.
In France in 1494, for example, a young pig entered a house and,
according to court records, "ate the face and neck" of a young
boy, killing him. The pig was jailed, tried, convicted, and hanged.

The earliest medieval animal trials date to thirteenth-
century Burgundy, and thereafter they spread throughout France
and into the Low Countries, Germany, and Italy. There was
biblical precedent for punishing animals. "When an ox gores a
man or a woman to death, the ox shall be stoned," according
to the book of Exodus, "but the owner of the ox shall be clear."

Oxen, however, rarely found themselves in the docket. Medieval courts disproportionately prosecuted pigs.

Pigs accounted for well over half of all the animal executions in medieval Europe. Roaming free on farms and in towns, they sometimes fell victim to their own omnivorous appetites. In modern-day Papua New Guinea, where pigs wander through villages much as they did in medieval Europe, children are sometimes bitten as the animals steal food from their hands. These medieval pigs were perhaps attracted by bit of gruel on an infant's upper lip and then got carried away.

The owners of killer pigs were held blameless. When a young pig was hanged for killing a five-year-old boy near Chartres in 1499, the pig's owners were fined—not for failing to control their pig but for failing to protect the child, who had been left in their care. The guilt for the murder itself lay entirely with the animal.

To modern minds, the rationale for such trials seems bewildering. One European court explained that a pig would be hanged so that "an example may be made and justice maintained," as if other pigs might heed the lesson. In another case the court noted that the pig had killed an infant and eaten its flesh "although it was Friday": the animal, in other words, had violated not only the commandment against murder but also the church's prohibition against eating meat on that day of the week.

A particularly unusual execution took place in France in 1386, after a sow killed a three-year-old boy. The animal was dressed in a jacket and trousers, with white gloves on its front hooves and a mask resembling a human face over its snout, and hanged not by the neck but by the rear feet. Local laws mandated the upside-down position, if not the costuming: "If an ox or horse commit one or more homicides," the law noted, the beast should be forfeited to the local lord but not killed. "But if another animal or a Jew do it, they should be hung by their rear legs." Pigs and Jews suffered the same sort of humiliating execution.

Pigs loom large in the appalling history of European anti-Semitism. People are often defined by their foods: Englishman are roast beefs; Sicilians are macaronis; the French are frogs. Jews, in an odd reversal, became most closely identified with the animal they refused to eat. English illustrations of the crucifixion often depicted Christ's tormentors as humans with pig snouts, just one expression of the familiar charge that the Jewish people were "Christ killers." In Germany a common anti-Semitic image was the *Judensau*, or "Jew's sow," which portrays Jews suckling at the teats of a giant sow and eating her excrement—a vicious echo of Romulus and Remus being suckled by a she-wolf. Martin Luther, in a religious tract, addressed Jews directly: "You are not worthy of looking at the outside of the Bible, much less of reading it. You should read only the bible that is found under the sow's tail, and eat and drink the letters that drop there."

Given the pig's diet, Jewish abstention from pork seems wise: pigs, after all, killed children, scavenged corpses, and ate feces. The great rabbi Maimonides, writing in the twelfth century, linked the biblical pork prohibition to the animal's habits. Jews rejected the pig because of "its being very dirty and feeding on dirty things," he explained. "If swine were used for food, marketplaces and even houses would have been dirtier than latrines." Christians, who cooked and ate these nasty beasts, might reasonably be considered more piglike than Jews.

But there was no problem of logic that could not be solved by fancy theorizing. Medical practice at the time derived from the Greek physician Galen's theory that good health required a balance among the body's four "humors": blood, phlegm,

Medieval Europeans tried and executed dozens of pigs for the crime of killing children. The sow depicted in this illustration was hanged by her rear legs, a humiliating measure reserved for pigs and Jews. The vicious anti-Semitism of medieval Europe often paired Jews with swine, perversely equating Jews with the animal they refused to eat.

choler, and melancholy. Food could be used to adjust a disordered system, but not in any simple way: for instance, chicken, being delicate, improved the humors of those with weak constitutions but might incinerate completely within the bodies of the strong, leading to "burnt" humors. Digestion, physicians believed, was a complex process that transformed foreign matter into human flesh, and the foods easiest to digest were those most similar to the human body. This led to the troubling conclusion that cannibalism was a wise dietary choice. According to one authority, no other food "is more agreeable to man's nourishment than human flesh."

Human flesh being forbidden, pork offered the best substitute. Galen had first suggested the similarity between the two. The idea emerged, perhaps, from the similar diets of people and pigs or from their similar anatomy. At a time when the church forbade human dissection, pigs served as substitutes. The twelfth-century text *Anatomia porci* advised dissecting pigs because the body of no other animal "appears to be more like ours than is that of the pig." The similarities, some said, were culinary as well as anatomical. One medical book reported, "Many have eaten man's flesh instead of pork, and could perceive neither by the savour nor the taste but that it had been pork." A butcher reportedly passed off human flesh as pork until one unlucky diner found a finger in his meat.

Because of its similarity to human flesh, pork was considered the most healthful meat—but only for Christians. According to the theory of humors, when people ate an animal, they could absorb its behavior along with its flesh. One might assume that this would render the eaters of pigs more piglike, but it wasn't that simple. As Christians saw it, both Jews and pigs were prone to lust and gluttony. Jews would grow even more sinful if they ate an animal afflicted with those same qualities; God, the wise physician, had therefore forbidden them to eat it. Christians saw themselves as in better control of their sinful natures and therefore capable of enjoying the benefits of pork without its drawbacks. Christians, one authority explained, can transform even dangerous food into virtuous nourishment, "just as honey changes the bitterness of the orange's peel into sweetness."

It gets worse: some Christians asserted that Jews, denied the meat of the pig, lusted after its closest equivalent: human flesh. An English rhyme told the tale of Hugh of Lincoln, an eight-year-old Christian boy supposedly killed by a Jewish woman.

> She'd laid him on the dressing table,
> And stickit him like a swine.
> And first came out the thick, thick blood,
> And syne came out the thin.

European Christians learned hatred of Jews in the cradle, through nursery rhymes and legends, and nothing was more frightening than tales of children killed by Jews, their flesh salted and eaten, their blood collected and used to make matzo or to concoct magical potions. These invented tales had brutally real effects: Jews were tried and executed—often hanged upside down, like child-killing swine—for allegedly committing these crimes. Such "blood libel" accusations, fantastical as they may seem, were taken seriously down through the Nazi era and persist even today.

Perhaps these theories involving pigs and Jews were just elaborate post hoc justifications for popular prejudice. The connection might have been as simple as this: pigs were the most despised animals, and Jews were the most despised people. A London town ordinance of 1419 referred to "Jews, Lepers, and Swine that are to be removed from the City." All threatened Christians with filth and contagion.

Equating pigs with Jews didn't stop some Christians from embracing pork as a symbol of their faith. This was especially true in Spain. The Visigoth rulers, who adopted Christianity in

589 AD, passed laws promoting the raising of pigs. Monasteries kept large herds of swine, and in many towns the central religious festivals involved Saint Anthony and Saint Martin of Tours, both closely associated with pigs. Unlike much of the rest of Europe, Spanish farmers continued to raise pigs on acorns in the forests and thereby preserved the pig's reputation as a noble creature of the woods rather than a dirty scavenger of the streets.

Muslim forces invaded the Iberian Peninsula in 711 and ruled over it for nearly eight hundred years. According to the Quran, "Forbidden to you is that which dies of itself, and blood, and the flesh of swine." All of those substances were also prohibited to Jews, and the Islamic law clearly owed a debt to its fellow Abrahamic religion. Environmental and political reasons—the unsuitability of swine for arid conditions and the desire to prevent the poor from raising their own food—likely also played a role in the Islamic pork ban. As Islam spread rapidly across the globe, it arrived in regions like Spain, where pigs were central to agriculture and cuisine. During the centuries of Muslim rule, Catholic monasteries protected Spain's legacy of swine.

After the Reconquista—the retaking of Spain by Christian forces, completed about 1500—pigs emerged from their sanctuary among the monks and assumed a prominent role. The Christian authorities carried out a policy of forced conversion of Muslims and Jews, with an emphasis on questions of diet. Christians assumed that their new coreligionists would rejoice at being freed from the burdens of the dietary laws—after all, Christians had long assumed that Jews secretly craved pork. One Christian text depicts Jews lamenting the culinary pleasures they had denied themselves and crying out, "How much ham we could have had!"

This coerced conversion made Christians fear—not without reason—that erstwhile Muslims and Jews were secretly maintaining their former ways. Many converts tried to combat such suspicions by displaying in their homes a slice of pork, called a *medalla* or medallion, as a fleshy talisman to ward off the Inquisition. In a work by the great playwright Lope de Vega, a character explains that he hung a side of bacon on his wall "so that the King will know that I am neither a Moor nor a Jew."

Inquisitors became obsessed with pork consumption. A convert named Gonzolo Perez Jarada appeared before the Inquisition in Toledo in 1489 to answer charges that he "did not eat bacon." In a similar case, a woman named Elvira del Campo was tried in Toledo in 1567 on charges of "not eating pork." The official record states that she was stripped, put on the rack, and interrogated. "I did not eat pork for it made me sick," she said at first. Then, after cords were twisted tightly around her wrists—"They hurt me! Oh my arms, my arms!"—she confessed that she abstained from pork because she remained an observant Jew. The Inquisitors confiscated her property and sentenced her to three years in prison.

Following the Reconquista, eating pork—even more than partaking of the Eucharist at a Catholic Mass—became the key marker dividing Christian from Jew. At about the same time, pork consumption emerged as a different type of boundary marker as well, one that delineated not just religious groups but also social classes. The demographic swings of the fourteenth and fifteenth centuries—a human population boom, followed by a fearsome die-off due to famine and disease—brought surprising changes to the way pigs lived and the way people thought about pork.

EIGHT

"THE HUSBANDMAN'S BEST SCAVENGER"

In Sir Walter Scott's *Ivanhoe*, a jester asks a swineherd what he calls the animals under his care. "Swine, fool, swine," the herdsman replies. The jester then asks what the animal is called after it is butchered. "Pork," comes the reply. The jester, after playing dumb, then makes an astute point about power, language, and food: The word "swine" is Saxon in origin, while "pork" is French. When living and under the care of "a Saxon slave," the animal goes by its Saxon name, the jester explains. When cooked and served at a "feast among the nobles," the swine magically becomes pork, because French is the language of the ruling class.

The scene dramatizes a great moment in linguistic history. *Ivanhoe*, published in 1820, is set in the twelfth century, immediately following William the Conqueror's conquest of England in 1066. When French nobles took over, their language took on a higher status, while the words of the defeated Saxons became vulgar. Thus "swineflesh" became pork (from the French *porc*), "cowflesh" became beef (*boeuf*), and "sheepflesh" became mutton (*mouton*). The scene in *Ivanhoe* points to a corresponding distinction in who ate what: "slaves" cared for livestock, but "nobles" ate them.

Such distinctions grew more significant over time. Tribal cultures, such as the very first farming villages of the Near East, had produced little wealth and therefore had simple social hierarchies consisting of food producers and rulers. Their cuisines had remained undifferentiated: kings and farmers ate the same foods. But as economies grew richer, societies were minutely carved into many classes—soldiers and laborers, priests and merchants, peasants and nobles. Diet helped define status. This had been true in ancient Mesopotamia, where the priests dined on lamb and the laborers ate pork, as well as in Rome, where senators banqueted on suckling pig and slaves made do with bread and tripe.

In medieval and Renaissance Europe, pork came to define status, but in complicated ways. At first a ubiquitous food for all, pork later became a luxury enjoyed only by the rich, then, later still, a dangerous food fit only for peasants. Pigs, meanwhile, continued to lose habitat and food sources: forests fell to the axe, commons were enclosed, and more cities banished free-ranging animals. By 1600, the European pig was in serious decline—until it found a new niche within modern commerce and received an infusion of fresh blood from China.

B etween 1000 and 1400 AD, Europe's population and food supply experienced wild swings. Before 1000, when Europe's

forests were deep, pigs ran wild and nearly everyone ate pork—
the only key distinction being that the rich ate more of it than the
poor. Such widespread meat eating was possible only when the
human population remained small and rangeland for livestock
was plentiful. When the Medieval Warm Period started about
950 AD, higher temperatures lengthened the growing season.
This prompted farmers to clear forests and plant more crops,
and the human population grew in tandem with the supply of
grain: between 1000 and 1350, the number of people in Europe
exploded from about 25 million to about 60 million. More food
meant more people, and more people meant more demand for
food. As the great economist Thomas Malthus would later explain, this was bound to end badly.

The population boom changed the types of food people ate
because eating plants is more efficient than eating animals: on
a given plot of land, growing grain can produce twenty times
as many calories as raising livestock. As the number of humans
rose in Europe, the number of farm animals plummeted. Nobles continued to eat large amounts of flesh—as much as three
pounds of meat and fish per day—but the peasant diet consisted
almost entirely of cereals, which lacked protein and essential nutrients. As this trend continued, the ecological and health effects
became more severe. By 1250, intensive cropping had drained
the soil of nutrients, and there were fewer farm animals producing manure for fertilizer. Already malnourished, European
peasants began to starve. Soon environmental change would
make the situation even worse.

Around the turn of the fourteenth century, the Medieval
Warm Period gave way to the Little Ice Age, and the colder
weather hurt harvests. By 1300, Europe had seen the first of
a series of crop failures and famines that would devastate the
region over the next half century. When the Black Death—most

likely bubonic plague carried by fleas on rats—arrived in southern Italy in 1347, it found a continent weakened by famine and unable to fight off disease. Over the next four years, about a third of the people in Europe died.

The Black Death produced one positive side effect: peasant diets improved. Demand for food fell, which caused prices to decline. In France and Germany the price of grain plunged by as much as 70 percent. With workers scarce, wages rose and peasants could afford to buy meat. In 1397 the average resident of Berlin ate more than three pounds of meat a day, far more than today. On one manor in Norfolk, England, harvest workers in the pre-plague era had received just one ounce of meat with every two pounds of bread; after the plague, a full pound of meat accompanied those two pounds of bread. Europeans may have suffered the horror of watching a third of their neighbors die, but they at least could console themselves with a good meal.

The Black Death democratized meat, and democracy is always troubling to the elite. When the meals served in castles began to resemble those cooked in cottages, Europe's nobility made a dietary pivot, and pigs were one of the animals most affected. During the prior few hundred years, when peasants had supped on gruel, eating pork offered nobles sufficient proof of their elite status. Now, as pork became affordable even to the lower orders, the wealthy began to spurn it.

Europe's elites turned from hoofed livestock to winged beasts. When archaeologists dig up castle sites around England, they find that pig bones begin to dwindle not long after the Black Death and are replaced by those of fowl, especially wild birds, which had become the new marker of wealth. In 1501 the Duke of Buckingham hosted a meal that omitted pork, beef, and mutton entirely; instead he served five pheasants, twelve partridges, twenty-four chicks, six capons, twelve rabbits, and thirty-six small birds. Particularly suspect to Europe's wealthy were the cheapest and most widely available forms of pork: sausage and bacon. Even the finest sausages and other cured meats were fit only for merchants and the more affluent peasants. Woodcut illustrations of peasant weddings from this era almost invariably showed tables laden with sausages and a dog running off with a purloined strand.

Physicians gave these new prejudices the sheen of medical authority. A couple of thousand years of medical thought had promoted pork as the most healthful meat, but physicians in the sixteenth century revised that tradition. They now argued that pork, rather than being the easiest flesh to digest, was the most difficult: only peasants toiling in the fields produced sufficient heat in the body to break down pork. One Renaissance doctor advised that the sedentary elite should restrict themselves to lighter fare, while sausages and bacon were fit only for the "rustical stomach."

For reasons of status, health, or both, the elite avoided pork. "Pork is the habitual food of poor people," a visitor to Paris observed in 1557. A century later a Frenchman noted, "With the exception of hams and a few other more delicate portions, today only the lower classes are nourished on pork." In Scotland, another writer reported, "pork is generally despised, and left to be consumed by the mean populace."

The "mean populace," however, delighted in pork, especially on festival days when they gorged on roast pig and sausages. The sausage played the same role in the Renaissance as the hot dog in twentieth-century America—providing a cheap, filling meal for urban crowds. The elite, then as now, suspected that the common people congregated only to feast, gamble, and whore. Though the wealthy were prone to these vices as well, they could indulge privately. The poor, among their many misfortunes, were forced to sin in public, and their sinning and pork eating became intertwined.

The English sometimes referred to a brothel as a "hog house," and indeed the connection between pork and sex stretches back to the ancient world. The boys who sold sausages at Greek and Roman markets often doubled as prostitutes. And then there is the inescapable fact that chopped meat stuffed into an intestinal casing produces a rather suggestive shape. The most common Greek word for sausage, *allas*, makes its first written appearance referring not to food but to sex: a line from the Greek poet Hipponax describes an aroused man "drawing from the tip down, as if stroking a sausage."

The sexual associations of pork carried on into the Renaissance. Bartholomew Fair, a riotous event that took place in London each August, gave its name to the Bartholomew pig, roasted whole and served to fairgoers eager to indulge all of their sensual cravings. Shakespeare's Falstaff, a man of large and indelicate appetites, is described as a "whoreson little tidy Bartholomew boar-pig." In Ben Jonson's comedy *Bartholomew Fair*, the most dissolute characters gather at a pork-selling stand operated by Ursula, an obese, filthy, delightfully foul-mouthed "pig-woman." Zeal-of-the-Land Busy, a Puritan intent on remaining holy despite the fair's corrupt enticements, decides to sample a bit of pork and ends up devouring two entire pigs—proof, as another character suggests, that "the wicked Tempter" does his work through "the carnal provocations" of the pig.

Jonson was satirizing Puritan hypocrisy, but the fear he targeted was real enough. Pork, in the minds of many Renaissance Europeans, was for people who had fallen prey to fleshly appetites. Those who wished to maintain control—Puritans, nobles, priests, a rising bourgeoisie—had best choose other foods.

Avoiding pork was getting easier. After the Black Death, abandoned fields didn't return to forest but instead became pasture for sheep and cattle. With grazing more abundant, England's romance with roast beef came into full flower. The wool industry boomed, producing abundant cheap mutton as a by-product. The enclosure movement, which brought land previously held in common under private control, eliminated the woods and wastes where peasants had formerly kept pigs. By 1696, England had about 12 million sheep and 4.5 million cows, but only 2 million pigs.

Most farmers kept just one or two pigs to convert waste. Gervase Markham, in a 1614 book titled *Cheape and Good Husbandry*, judged pigs "the husbandman's best scavenger, and the huswive's most wholesome sink," because they eat everything that would otherwise "rot in the yard [and] make it beastly." In *The Wealth of Nations*, Adam Smith praised the pig because it "greedily devours many things rejected by every other useful animal" and as a result can be "reared at little or no expense." This trait made the pig attractive, but only in small numbers, limited by the amount of available waste.

By the seventeenth century, however, a growing economy had created a new niche for pigs. Activities such as dairying and bread making, once undertaken in every household, became large commercial enterprises. The concentration of by-products rose, and so did the concentration of pigs: they began to devour all sorts of commercial wastes. One writer noted that pigs could be fed "chandlers' grains," the crispy bits of flesh and gristle left over after rendering beef fat into tallow. In 1621 a London

By 1700 pigs were far less numerous in England than cattle and sheep, ruminants that provided milk, wool, and more highly prized meat. Most farmers kept only a few pigs to eat agricultural waste, as in this illustration from a 1732 farming guide. Soon, however, pigs would be raised on a larger scale to eat the by-products from commercial dairies, breweries, and distilleries.

maker of starch—refined wheat used to stiffen clothes—fattened two hundred pigs on his leftover bran. Alcohol production provided an even larger source of feed: thousands of pigs lived in lots adjacent to distilleries and breweries to consume the spent grains. At dairies, milk cows shared space with pigs. Daniel Defoe reported that Wiltshire and Gloucestershire produced "the best bacon in England" from hogs fattened on "the vast quantity of whey, and skim'd milk . . . which must otherwise be thrown away." Dairymaids churned butter, and the whey flowed through a channel directly to the pig trough. One writer defined a dairy as "a center about which a crowd of pigs was collected."

These large herds of swine found eager buyers. In the great age of exploration, sailors needed foods that wouldn't spoil during long voyages. This prompted a vast expansion of the salt-food industry, as pork, beef, and fish were packed into barrels and rolled aboard ships. Pork, because it preserved so well, commanded much of this market. The British navy required as many as 40,000 pigs annually, and a member of the Victualling Commission explained that he bought mostly "town-fed hogs" fattened in the yards of liquor distillers.

By the later eighteenth century, the navy had found another source of pigs. As part of the so-called agricultural revolution, farmers had started to employ new crop rotations that involved planting peas and beans to fix nitrogen and revive exhausted soil. Those legumes became hog feed, and farmers sold the pork to the navy.

Thanks to these abundant new food sources—legumes, dairy waste, distillery grains—farmers could raise pigs on a larger scale, a circumstance that favored changes in the pig's constitution. Swine no longer roamed the streets or the woods to find their own food. Now they lived in pens and had their food delivered to them. Farmers suddenly had a strong interest in determining which types of pigs turned feed into meat most efficiently.

Like Rome a couple of thousand years earlier, London had created conditions that favored a fat sty pig. The first step in this direction was the Old English hog, a breed of obscure origin that existed in slightly different varieties all over the country. Unlike the prick-eared medieval pig, the Old English pig had lop ears drooping over its eyes. Rather than black or brown, it tended to be white, mottled, or saddled. In many cases, these pigs reached slaughter weight at about eighteen months—older than the six months of modern pigs but younger than the two or three years common for woods hogs.

The Old English marked a small improvement over the medieval forest pig, but a bigger change was on the horizon. An English agricultural writer picked up on this in 1727 when he noted the recent appearance of some odd pigs, "the little black sort with great bellies." The animal he describes sounds nothing like any European variety—but very much like a pig from China.

Chinese swine began infiltrating Europe about the same time that the industrial and agricultural revolutions picked up steam. Analysis of mitochondrial DNA shows that European and Chinese pigs first started swapping genes sometime in the eighteenth century. And Chinese hogs were, at this point, far better suited than their European counterparts to the conditions of English agriculture.

China, over the centuries, had developed a devotion to pigs just as intense as that of ancient Rome or early medieval Europe. In Neolithic China swine had served as a key source of wealth, and in the second millennium BC, they were commonly used in

Unlike their forest-dwelling European counterparts, Chinese swine had evolved into fat creatures of the sty, used to produce both meat to eat and manure for fertilizer. Imported to Europe after 1700, Asian pigs—such as this one depicted in an 1858 American farming manual—were interbred with European varieties to create the modern pig breeds we know today.

sacrificial offerings. In the centuries to come, pork remained the daily meat of wealthy Chinese and the key animal protein in a sophisticated cuisine. For China's poor, as for Europe's, it was the food of festivals and of survival. Peasants marked the New Year with the slaughter of the family pig, consuming the organs immediately and selling the meat or salting it away for the coming year. Even in the twentieth century, pork accounted for 70 to 80 percent of the calories from animal products consumed by the Chinese.

Overall, however, pork represented just a tiny part of the Chinese diet: in the twentieth century it accounted for just 2 percent of the total number of calories consumed by the Chinese, compared to 83 percent from grains and 7 percent from legumes. And this is not a modern development. The ancient works on Chinese agriculture virtually ignore animal husbandry, concentrating instead on rice, millet, wheat, and soy production.

How, then, did the pig come to play such an outsized role within China's overwhelmingly vegetarian society? In the words of Chairman Mao, the pig was a "one-man fertilizer factory." Though a large country, China's terrain is largely rugged, with limited areas suitable for planting. It faced the problem of a growing population and dwindling food sources far earlier than Europe—as early as the third century BC—and responded with advanced agricultural methods. Pressure on the land was especially strong in the subtropical rice-growing regions of central and southern China, where every scrap of arable land was brought into production and farmers produced two or even three crops a year. This left no open land for pasturing sheep or cattle, and the intensive growing threatened to strip all nutrients from the soil. But the pig saved the day, ensuring that China's soil didn't become depleted the way that Europe's did.

Chinese swine were penned and fed on agricultural waste—especially the hulls of rice—mixed with wild-growing plants such as water hyacinth. The pig functioned as a composting machine, transforming coarse vegetation into precious fertilizer. More intensive agriculture increased the need for fertilizer, but it also boosted the amount of agricultural by-products to feed pigs and therefore the quantity of manure. Pigpens were constructed with watertight floors to collect not only feces but also urine, an especially rich source of nitrogen. When modernizers introduced American pig breeds into China in the 1930s, the farmers complained that though the animals grew quickly, they produced too little manure.

From a very early date, pigs in China were confined to small sties, and this created distinct evolutionary pressures.

They became short-legged, swaybacked, and potbellied, with squashed snouts and concave faces. They were adapted to eat, gain weight, and breed. Some Chinese sows produced litters of twenty piglets.

Merchant ships began plying the ocean routes between Europe and Asia in the sixteenth century, and by 1700 or so, they had carried Chinese pigs to England. The timing was propitious. In the European forests of 900 AD, the Chinese pig would have been easy prey for wolves and bears, and it would have lost badly to the forest pig in the race to gobble up falling acorns. But it was adapted well to the conditions of eighteenth-century industrial England: living in pens and eating beans, distillery grains, and dairy waste. The subtropical Chinese pigs were a bit delicate for colder British conditions, so breeders crossed them with European types, hoping to produce a hardy pig that would gain weight quickly and produce piglets by the dozen. Eventually, that is exactly what happened: the forest pig was pushed to the margins in Europe, ousted by its fatter rival.

But the forest pig's moment in history hadn't quite passed— it had new lands to conquer. The wild boar and domestic pig had spread throughout Eurasia, from Norway to Thailand, but oceans had prevented them from colonizing the Western Hemisphere. That changed after 1492, when ships started sailing west across the Atlantic. The New World, as it turned out, was a perfect place for pigs. Swine thrived in the Americas and played a crucial role in assisting Europeans as they conquered new lands.

NINE

"ALL THE MOUNTAINS SWARMED WITH THEM"

Spain's King Ferdinand and Queen Isabella had little faith in Christopher Columbus's plan to find the Orient by sailing west, so in 1492 they outfitted him grudgingly with three small ships and a crew of ninety men. When he returned to Spain in March the following year with a few gold trinkets, a flock of parrots, and a handful of captive "Indians," the monarchs' enthusiasm grew. Six months later Columbus departed again with seventeen ships holding 1,200 men and enough supplies to establish a permanent colony, including everything they needed to re-create the Spanish diet in the New World. They took seeds and cuttings for wheat, chickpeas, melons, onion, radishes, salad greens, sugarcane, grapes, peaches, pears, oranges, and lemons. The ships also carried a menagerie of domestic animals: chickens, horses, cows, sheep, goats, and pigs.

These voyages started what has become known as the "Columbian exchange." The landmasses of the Americas had been separated from those of Eurasia and Africa for millions of years, and the people of the New World had had no contact with those of the Old World since the first migrants crossed over the Bering Land Bridge into the Americas ten or twenty millennia earlier. Columbus and his fellow explorers changed all that by inaugurating the era of transatlantic travel, which transformed the biota of the entire world.

Plants and animals flowed in both directions across the Atlantic. In the realm of food, the New World contributed the potato, sweet potato, and corn, as well as chocolate and tomatoes. Within a few hundred years, these foods had reshaped the diets of people from Ireland to Africa to China, often for the better. The people of the New World, by contrast, experienced the Columbian exchange in a more immediate and catastrophic way. First, European epidemic disease, to which Indian peoples had no immunity, killed off much of the indigenous population. Then Spanish soldiers conquered the survivors and set about plundering the continent's gold and silver.

In conquest and colonization, the Spanish had a key ally that has earned little credit: the pig. An army travels on its stomach, and Spanish soldiers filled their bellies with pork.

Feeding the invading soldiers proved difficult at first. Columbus set up base on the north side of the island of Española, in what is now the Dominican Republic. Wheat and other European crops failed in the hot, wet climate, so Spaniards ate mostly cassava, the starchy root that was the dietary staple of the Taino, the island's native residents. Sheep, like wheat, wilted in the damp heat. Cattle showed more promise but would need a few generations to acclimate to the weather and build up a herd—time the conquerors didn't have.

Pigs never missed a beat. As soon as their cloven hooves landed in the soft jungle mud of the Caribbean islands, they started eating and breeding. Just two years after Columbus's second expedition made landfall, one Spaniard noted that the pigs "reproduced in a superlative manner." In a few more years the number of hogs running wild was *infinitos*, and "all the mountains swarmed with them."

Pigs are the weediest domestic animal—opportunistic, tough, and fecund. Like rats, they can live nearly anywhere; unlike rats, they taste good. Spanish and Portuguese sailors dumped breeding pairs of pigs on uninhabited islands. "A sow and a boar have been left to breed" on a certain island, one Spanish explorer told another in a letter. "Do not kill them. If there should be many, take those you need, but always leave some to breed, and also, on your way, leave a sow and a boar on the other islands." In 1609 the ship *Sea Venture*, on a mission to bring supplies to starving colonists at Jamestown, Virginia, was blown off course by a hurricane and ran aground on the shoals off Bermuda. When the 150 people aboard made it to shore, they were overjoyed to find plenty of pigs, descendants of a batch left by the Spanish a century before.

Pigs loved the warmth of Española, the dense forests, the coastal marshes, the plentiful rainfall, and the lack of predators. They especially loved its food. Columbus wrote that the trees and plants there were "as different from ours [in Europe] as day from night," but they were agreeable to pigs, which rooted up the Tainos' cassava and sweet potatoes and devoured their guavas and pineapples. They snatched baby birds out of nests and lizards and snakes from the ground. Their favorite food was the *jobo*, a plum-sized fruit from a tree native to the American tropics. One Spaniard risked blasphemy by claiming that pork from pigs fattened on this fruit tasted even better than the acorn-fed variety back home.

The Spaniards shipped pigs to other islands as well. The Jamaican mountains soon held what one witness called "countless herds," and in 1514 the governor of Cuba told King Ferdinand that the herd of 24 pigs he had carried to the island less than two decades before had ballooned to 30,000.

The European pigs had strolled into an empty niche. Before Columbus's expedition, the only mammals living in the Greater Antilles—the island chain that comprises Cuba, Hispaniola, Jamaica, and Puerto Rico—had been the Taino people, bats, and a largish rodent called the hutia. The mainland of North and South America hosted thousands of land mammals, but few were domestic: the peoples of the Andes had domesticated llamas, alpacas, and guinea pigs, and further north some tribes kept turkeys and ducks. From Patagonia to the Arctic, native peoples kept dogs, which had made the trip with the first humans to colonize the Americas. All of these beasts had their uses, but none could match Europe's livestock for the purposes of transport, traction, and meat production.

There's a simple reason for this disparity: the Americas contained few wild animals suitable for domestication. Peccaries, the American cousins to Eurasian pigs, are territorial and lack the strong dominance hierarchy vital to the process. Bison, though similar to cattle, are prone to stress disorders, while deer and moose are solitary and skittish. Two good candidates, horses and camels, had become extinct in North America around 11,000 BC, probably as a result of overhunting by humans. American societies had developed without the livestock that had contributed so much to the strength and wealth of European peoples.

That lack of livestock also left the Indians vulnerable to disease. For thousands of years Europeans had lived in dense communities thick with domestic animals. This formed the perfect breeding ground for pathogens, which spread from person to person and often jumped the species barrier between humans and animals, evolving new ways to kill. When Spaniards and their animals arrived in the Americas, they carried with them smallpox, measles, whooping cough, bubonic plague, malaria, yellow fever, diphtheria, amoebic dysentery, and influenza—along with the antibodies necessary to survive these scourges. Europeans, that is, had grown up with these diseases and therefore developed some immunity to them. The Indians had not. What followed is known as a "virgin-soil epidemic," when the deadly diseases of the Old World infected Indian peoples whose immune systems had no defenses against them.

The mortality rates of these epidemics made the Black Death seem relatively minor by comparison. In the century after Hernando Cortez invaded Mexico, the Indian population plummeted by as much as 90 percent. Disease swept through the Americas so quickly that most Indians encountered European pathogens long before they met any European people. Spanish soldiers, brutal as they were, merely administered the coup de grâce when they defeated Native Americans in battle.

The islands of Española and Cuba functioned as advance bases for military forces intent on conquering the American mainland and plundering gold and silver. Outfitting an army from Spain would have been nearly impossible, given the long ocean voyage to the New World. The Caribbean islands, with their rapidly breeding herds of animals, served as a Spanish commissariat. Within three decades of Columbus's first landing, the

Starting in 1492, corn, tomatoes, and potatoes were shipped to the Old World, and the full complement of European domestic plants and animals—as well as infectious diseases—arrived in the New World. Pigs, appearing at bottom left in this sixteenth-century illustration, played an unheralded but crucial role in the Spanish conquest of the Americas.

Spanish had assembled everything they needed to conquer most of the Western Hemisphere.

To assist its soldiers, Spain relied on horses, dogs, and pigs. The horses allowed speedy travel over great distances and also intimidated opponents, who had never before seen a horse, let alone a man on horseback. The dogs, similar to mastiffs and wolfhounds, attacked and killed the enemy. Pigs formed a mobile food supply, trailing along at the rear of the column. Horses and dogs found glory in battle. Pigs, cast in a supporting role, have been largely forgotten. A Spanish historian has argued that, though the horse contributed greatly to the conquest, "the hog was of greater importance and contributed to a degree that defies exaggeration."

Having bred so promiscuously on the islands, pigs were the cheapest source of meat available to the hordes of soldiers amassing in the Caribbean. Once ashore on the mainland, swine could be counted on to start breeding just as prolifically as they had on the islands. Unlike cows, they didn't require pastures but could pick up their living along the trail. The one supposed downside—the difficulty of herding pigs—was not a problem at all: pigs can be herded quite easily, as the Spaniards knew well.

The conquistadores Hernán Cortés, Francisco Pizarro, and Hernando De Soto hailed from Extremadura, a highland region in the west of Spain known for its pig herds. Legend has it that Pizarro, conqueror of Peru, was a bastard who had been abandoned as an infant and survived by suckling at a sow's teats. This last charge was plainly a slander, although Pizarro was, in fact, a bastard. His father was a member of the lesser nobility whose moderate wealth derived in part from hog raising; his mother was a peasant girl working as a convent maid when she became pregnant. Pizarro grew up with his mother's family, and, like most peasant boys in Extremadura, he probably tended hogs. His enemies called him a swineherd to insult him, but the barb carried little sting in his home region.

Since Roman times, writers have noted that the oak forests of Extremadura are especially suitable for pig grazing. The region is known, these days, as the home of *jamón ibérico de bellota*, or ham from acorn-fattened hogs. The key tree variety is the holm oak, which yields abundant crops of sweet acorns. On sandy soils closer to the coast, cork oaks provide acorns as well as bark to plug wine bottles.

Given the ample supply of mast for hogs, nearly everyone in Extremadura kept pigs: a traveler once observed that settlements in the region should be described not as villages but as "coalitions of pigsties." In 1554 one community in Extremadura reported that 100,000 pigs had been fattened there. The mountains also offered a climate ideal for curing pork, cooler than the lowlands in the summer and drier in the winter. By the time Columbus sailed for America, hams from the region had already become famous, and ships heading for the New World carried a stock of Extremaduran hams as well as live pigs.

Hernán Cortés, a native of the Extremaduran town of Medellín, counted heavily on pigs in his conquest of the Aztecs. Before his 1519 expedition, he purchased a large herd of swine in Cuba. Once on the mainland in Veracruz, he ordered one of his men to establish a pig farm to keep up the expedition's supply. After gaining control of Mexico, Cortés commanded that more pigs be brought from Española, Jamaica, and Cuba. And when he left Mexico to conquer Honduras, swine trailed behind the soldiers.

In 1531 Pizarro, a distant cousin of Cortés, loaded a herd of pigs aboard his ship and set sail from Panama to South America. He dropped off men and hogs on the island of Flores and later in the Peruvian mountains at Tumbes to establish breeding herds, then marched with the rest to conquer the Incan capital. In 1541 Pizarro's brother Gonzalo assembled a herd of more than 4,000 pigs in his expedition searching for the "Land of Cinnamon" east of Quito. Even that astonishing number wasn't enough: by the end of the journey, all the pork was gone, and the men resorted to eating their horses and dogs.

With South and Central America colonized by Spanish pigs, it was only a matter of time before North America followed. The invasion began in 1539, when Hernando De Soto, another Extremaduran, and his small army of Spanish soldiers stepped ashore in what is now Tampa, Florida. De Soto, who had helped defeat the Incas, had persuaded King Charles V to let him search North America for riches; he felt certain he would discover mines of gold and silver to rival those in Mexico and the Andes. He brought six hundred soldiers, two hundred horses, and thirteen pigs. Compared to Gonzalo Pizarro's enormous swine herd, that was a pitiful number. But De Soto took good care of his baker's dozen, the very first representatives of *Sus scrofa domesticus* to reach North America. He and his men marched at the pace of their pigs, about twelve miles a day. The expedition covered hundreds of miles through what is now Georgia, the Carolinas, Tennessee, Alabama, Mississippi, and Arkansas. When they encountered rivers, the Spaniards built rafts to ferry the pigs across.

De Soto rarely allowed his pigs to be eaten, viewing them as the core of a breeding herd required for permanent colonization. The soldiers primarily ate corn confiscated from the Indians, sometimes going for months at a time without meat. Only in dire circumstances would the leader grudgingly allow a pig to be killed: his chronicler recounts one such instance, when, having gone three or four days without corn, De Soto "ordered half a pound of flesh to be given to each man."

The pig coddling worked. When De Soto died of illness along the Mississippi River in 1542, his property consisted of four slaves, three horses, and seven hundred pigs—an increase of more than 5,000 percent in less than four years during an arduous journey across mountains and through swamps.

De Soto's expedition didn't find the precious metals it had sought, so Spain devoted few resources north of Florida. In Central and South America, though, the colonial enterprise thrived, with pork as the main food for Europeans. The Spanish had learned that colonization didn't work without pigs. The first settlement of Buenos Aires proved disastrous, in part because it lacked a foundation of livestock. That's why the crown, when granting licenses to settlers, required that their ships transport as many as five hundred pigs each. In Peru, vast amounts of pork were required at Potosi to feed the miners, who labored under horrendous conditions and lived scarcely longer than the pigs driven to the mines to feed them. Masters as well as slaves ate pork. Toluca, in the highlands of central Mexico, developed a reputation for making hams and sausages nearly as delicious as those from Extremadura.

Pigs proved essential to the exploration, conquest, and initial colonization of Latin America, but their heyday lasted less than a century. By 1600 beef had replaced pork as the dominant meat among colonists in the Americas. Cows and sheep bred more slowly than pigs, but over time they built up herds and provided wool and hides in addition to meat. Compared to pigs, both were better equipped by nature to deal with the hot, dry climates found in much of Central and South America. Spanish cattle, ancestors of the tough Texas longhorn, adapted to the grassy plains of central Mexico and the vast pampas of Argentina, a resource the Indians had rarely used because they kept no large grazing animals. Sheep, which needed even less moisture than cattle, thrived in the Gran Chaco, the lowland plain that extends through Argentina, Paraguay, and Bolivia, as well as in the highlands of the Andes and Mexico. In most places pigs became village scavengers, food for peasants, just as they had been back in Europe.

Colonial economics hastened the change. Hogs were best suited to the warm, wet tropics and subtropics of the Caribbean and Brazil, but those lands were soon dedicated to a crop far more lucrative than pork: sugarcane. Hogs became so scarce that, to feed their slaves, sugar planters had to import salted pork from a new source, the English colonies in North America.

Spain's settlement of Latin America had been a project of conquest, with pork as the key military food supply. In New England and Virginia, by contrast, the typical European was not a soldier but a farmer. Even so, the pig became even more essential in the English colonies than it had been for the Spanish. Pork served as food for colonists and a commodity for export, while the pigs themselves became a pestilence, helping to drive away the native peoples and clear the land for the English.

TEN

"A GREAT UNKINDNESS FOR OUR SWINE"

The English arrived late to the game of empire. By the time Sir Walter Raleigh landed on Roanoke Island in 1585, Spain had ruled the Caribbean for nearly a century. Back home in England, many suggested that New World colonists should concentrate on farming, but Raleigh treated agriculture with contempt. Spain derived its power from "Indian gold," he explained, not from "sacks of Seville oranges." Raleigh planned to follow the example of the Spanish, who colonized with the single-minded purpose of mining gold and silver.

Raleigh, however, found no precious metals along the eastern seaboard of North America; nor did other English explorers. So they chose to make a virtue of necessity. The English began to describe the New World's gold and silver mines as a "poisoned chalice" that had corrupted the Spanish soul. Under cover of papal authority, the Spanish had stolen the land and riches of the native peoples. The English, good Protestants, vowed to operate differently. They would be the anti-Spanish—farmers rather than soldiers. Instead of mining for gold, they would grow crops. Instead of killing Indians, they would convert them to Christianity and train them as farmers.

This posture nicely suited the Englishman's idea of himself. Neat fields, tidy hedgerows, healthy animals—all spoke to his virtue and godliness. Farmers needed to follow daily and seasonal schedules, plan for the future, and save against the unknown. They maintained fences and hedgerows to show that their land had been wrested from a state of nature and brought under human control. The best symbol of the connection between agriculture and virtue, however, was the domestic animal, a creature that had been civilized and made to serve man's needs. The English saw themselves as fulfilling God's decree in Genesis: "Let them have dominion over the fish of the sea, and over the fowl of the air, and over the cattle, and over all the earth."

In one sense, the plan worked. Britain built an empire to rival Spain's, and its colonies in North America became among the wealthiest societies the world has ever known, with an economy built almost entirely on agricultural goods.

Other aspects of the plan met with less success. Rather than becoming tidy farmers on the English model, the colonists raised crops and livestock under such sloppy conditions that visitors from England were appalled. And Indians ultimately fared little better in England's colonies than they had in Spain's: those who didn't die of disease were forced off their land to make way for settlers, crops, and livestock. In this sense, pigs became agents of empire in their own right: they wandered the woods, devouring all available wild foods, and thereby helped destroy the Indians' way of life.

Before Europeans arrived, the native peoples of North America had changed the landscape to suit their needs. Some sixty years before Raleigh founded his colony at Roanoke, Italian explorer Giovanni da Verrazzano explored the coastal areas of North America and discovered a forest so clear of undergrowth that it could be traversed "even by a large army." He didn't understand that Indians had burned the shrubs and small trees to encourage the growth of grasses to feed deer. Native Americans had also cultivated the trees they liked best, including chestnut, white oak, pecan, walnut, beech, butternut, honey locust, mulberry, persimmon, and plum. They made milk from hickory nuts and flour from acorns and ate many other types of nuts raw or roasted. Europeans marveled at the productivity of the American forest; they had no idea it was really an orchard.

In the river valleys, Native Americans planted corn, the most important crop throughout the Americas. Columbus had been the first European to describe a tall grass with seeds "affixed by nature in a wondrous manner and in form and size like garden peas." First domesticated in Mexico, *Zea mays* was grown by Native Americans as far north as Montreal and as far south as Santiago. Tisquantum, a member of the Patuxet tribe who was better known as Squanto, taught the English to plant corn in 1621, and both the Pilgrims and later colonists would have starved without it.

The English settlers admired some of the Indians' agricultural habits. William Wood, in Massachusetts, praised the Indian women for keeping corn "so clear with their clamshell hoes as if it were a garden rather than a corn field." Mostly, though, the British criticized the Indians because they built no fences, raised no barns, and abandoned exhausted fields rather than fertilizing the soil. The Indians, wrote John Winthrop, first governor of the Massachusetts Bay Colony, "inclose no land"; nor did they have "tame cattle to improve the land by." Robert Gray wrote that in Virginia the "savages have no particular propriety in any part or parcel of that Country, but only a general residency there, as wild beasts have in the forest." As the Englishmen saw it, the Indians had no true ownership of the lands they had inhabited for thousands of years.

To justify seizing native land, the English adopted a Roman legal principal known as *res nullius*, "empty things," in which all land was common property until it was "improved," generally through agriculture—and agriculture of the European, not the Native American, variety. Winthrop noted that the colonists "appropriated certain parcels of ground by inclosing and peculiar manurance." That is, they fenced the land, fertilized it with manure, and therefore came to own it. In this way, the English set themselves apart from their imperial rivals. The Spanish claimed ownership over their American land because the pope had given it to them. The English, by contrast, would earn their possessions by bringing unused ground under cultivation. And, as they saw it, America was nothing but unused ground.

There was a good reason America seemed so untended, so empty: most of the native people had died of diseases brought by Hernando De Soto, Sir Francis Drake, and others. In New England an epidemic—perhaps hepatitis A—that started in 1616 killed nearly all of the coastal Indians. The Pilgrims, arriving in 1621 and seeing signs of the recent die-off, assumed that divine providence had removed the Indians to clear the path for Christians.

The colonists at first vowed to bring surviving Indians within the circle of society. They would do so, in part, by turning them into cattle herders. According to Roger Williams, founder of Rhode Island, keeping livestock would help Indians advance "from barbarism to civility." In 1656, Virginia's legislators offered Native Americans a bounty of one cow in exchange for eight wolves' heads, explaining that owning livestock was "a step to civilizing them and making them Christians."

Only cattle, the English believed, served this civilizing function. Horses encouraged mobility and were useful in warfare, which ran counter to the British desire to render Indians peaceful and sedentary. Pigs were self-sufficient, requiring little care. The keeper of a cow, on the other hand, had to secure pasture in summer and save hay for winter. Dairy cows performed this civilizing function even more dramatically, because daily milking required owners to maintain steady habits. For these reasons, cows became a symbol of civilization. On early maps of New England, the English used a cow icon to represent areas that had been brought under their control. As historian Virginia DeJohn Anderson has phrased it, colonists hoped that the Indians would be "domesticated by their own cows." That plan failed spectacularly. In fact, the English began to farm like barbarians.

In the New World even the English couldn't maintain English standards. The British Isles were land poor and labor rich, whereas the American colonies had the opposite problem. Europe in the seventeenth century had no shortage of people willing to work for room, board, and modest wages. American farmers, by contrast, found few laborers for hire as they undertook the work of clearing forests, planting crops, building homes, and repairing roads. Back home, the English practiced intensive agriculture, squeezing enormous production from tiny plots. In America, colonists had no choice but to turn to extensive farming, relying on vast tracts of land to supply their needs. Rather than tending their cows and pigs, they turned the animals loose to fend for themselves. Rather than fertilize exhausted soil, they cleared new fields. In other words, they began to farm like Indians.

A few New England towns looked like the mother country, with cows grazing on town commons, overseen by communally hired herders. But most farms told a different story: with no herders available, animals roamed the woods, finding food where they could and defending themselves against predators—or not—when the necessity arose. The same held true in Virginia, where colonists single-mindedly grew tobacco and treated food production as an afterthought. Indentured servants and slaves earned their keep by tending the cash crop, not by herding animals. The barns, cowsheds, pigsties, and dairies considered a necessity for a proper English farm were extravagances in Virginia, so the animals lived without shelter even in the winter.

Not all livestock fared equally well. Goats had the toughness to survive but also had the unfortunate habit of eating the bark off young fruit trees—a capital offense at a time when homemade cider and brandy provided the only sources of alcohol. Sheep, because of what one colonist called the "humility of their nature," made easy prey for wolves. Tending dairy cows was considered women's work, and more than 80 percent of servants were men. In any case, the available forage was low in quality, so a cow that would give two gallons in England might give two quarts in Virginia, making it hardly worth the trouble. Beef cattle enjoyed more success: given enough space, they found sufficient forage in forests and marshes, though they had a tendency to become mired in swamps.

Only hogs truly thrived—but conditions required a particular type. Back in England, farmers had imported Chinese stock and developed a fat pig suitable for life in the sty. But there were few sties in America. Colonists imported pigs from the Caribbean rather than England, which meant lower transport costs and animals better suited to American conditions, descended as they were from the wily forest pigs of Spain. England's agricultural elite lamented the "degeneracy of the American pig," but the colonists got exactly the swine they needed. "The real American hog," one observer said, is "long in the leg, narrow on the back, short in the body, flat on the sides, with a long snout. You may as well think of stopping a crow as those hogs."

The American pig was the same beast that had ranged Europe for thousands of years and helped Spain conquer Latin America. Roger Williams once saw a wolf kill a deer, then watched as two sows drove off the wolf and ate the deer themselves. That was the toughness needed to settle a continent.

Williams's story notwithstanding, venison formed a relatively small portion of the swine diet. On the ocean shores pigs raided oyster banks and clam beds. In the woods they found wild peas, vetches, roots, and mushrooms. In Carolina and Virginia they roamed the orchards for windfall peaches, devouring the flesh, then deftly cracking the stone to get at the kernel inside. Pigs were said to be particularly fond of snakes, holding them down with one hoof, administering a killing bite, then sucking them down like noodles. Most of all, pigs ate nuts that fell from trees. The chestnut and oak trees once cultivated by Native Americans now provided food for pigs, a bounty unimagined in the forests of Eurasia.

Pigs and cows had the run of the land because North America remained mostly "unimproved," despite colonists' best efforts to claim land through the *res nullius* principle. We associate open-range ranching with the American West, but at one time or another it was standard practice from coast to coast. England had operated under a similar system before 1000 AD, when populations were small and forests vast, but the landscape had long since been tamed. Not so in America.

SOUTHERN PINE WOODS HOG

WESTERN BEECH NUT HOG

As English farmers turned to Chinese-European hybrids suited to life in the sty, Americans stuck with pigs (above) similar to those that had thrived in European forests for centuries—tough creatures that fought off wolves, fed on acorns and roots, and provided a nearly free source of meat for colonists and pioneers too busy for careful methods of farming.

Legal principles in America took shape around these new arrangements. If a pig in England rooted up a wheat field, then the animal's owner paid damages. In America, the law demanded that crops, rather than animals, be fenced in, and animals could wander wherever they pleased. Custom demanded that a fence be "horse high, bull strong, and pig tight": four or five feet tall to keep a horse from jumping it, strong enough to prevent a bull from knocking it down, and solid near the bottom to keep pigs from going under or through it. If a farmer built fences to these standards and kept them in good repair, then he could seek payment from the owner of a beast that breached his barriers and damaged his crops. Keepers of inferior fences had no redress at all. Livestock had the legal right to all land, public or private, not protected by a proper fence.

Lackadaisical farming was a rational response to prevailing conditions. America's farmers were "the most negligent, ignorant set of men in the world," according to one English visitor. A more acute observer explained that the farmers had little incentive to improve their practices because "nature has been so profusely bountiful in bestowing the mediums of makeshift." When makeshift solutions proved so effective—when pigs fed themselves and multiplied beyond reckoning—why trouble with fussy English practices?

The quality of the animals suffered from this hands-off approach, but the quantity did not. Reports from North America echoed those from the Caribbean two hundred years before. One man in Virginia reported "infinite hogs in herds all over the woods." A planter in Georgia explained why farmers celebrated the pig: "They who begin only with a sow or two, in a few years are masters of fourscore, or a hundred head." Virginian Robert Beverley noted, "Hogs swarm like vermin upon the earth [and] find their own support in the woods, without any care of the owner."

There was money in all those hogs. In 1660, Samuel Maverick reported that "many thousand" cows and hogs were being killed "to supply Newfoundland, Barbados, Jamaica." William Pynchon of Springfield, Massachusetts, also packed huge numbers of hogs to supply the lucrative West Indian market. As a Barbados planter explained to John Winthrop, Caribbean plantation owners "had rather buy food at very dear rates than produce it by labor, so infinite is the profit of sugar works." Cured meat became New England's second-most important commodity for export, trailing only fish.

Salt pork, along with salt cod, provided New England with what it so desperately needed: a cash crop for export. The region's economy expanded quickly as a result of agricultural assets, and in this New England was not alone. By the second half of the eighteenth century, North America had become an economic powerhouse. While northern ports exported protein by the barrel, Virginia sold tobacco, and the Carolinas and Georgia grew cotton and rice. On the eve of the American Revolution, colonists enjoyed the highest standard of living in the world.

Native Americans did not share in the prosperity. Not surprisingly, they resisted British attempts to "civilize" them. They rejected cows for the same reasons the British recommended them: the beasts demanded too much attention. (Indians also tended to be lactose intolerant, which rendered dairy cows unappealing.) They didn't want to transform their lifestyle to meet British expectations; they wanted to maintain their own cultures. And they embraced only those aspects of European farming that conformed to their own ways. What they wanted was meat and fat, easily acquired. What they wanted, in other words, was pigs.

By the 1660s many Indian tribes both in New England and in the Chesapeake region had acquired hogs. A Choptico chief described a method of stock acquisition that was likely standard for Indians and Europeans alike: he came across a semiferal sow that had recently farrowed, killed her, and claimed the piglets as his own. Free-ranging pigs could be tracked and killed in the woods just like deer. In their villages, Indians threw scraps and garbage to pigs, much as they did with their dogs.

As deer and other wild animals began to disappear, pigs took their place—both in the North American wilderness and in the diets of Native Americans. Indians boiled hog carcasses to render fat just as they had once done with bears, and they used lard in place of bear grease to oil their hair and skin. When deerskins for making moccasins were in short supply, pig hides sufficed.

Whereas colonists had encouraged Indians to keep cows, hoping the savages would adopt British culture, the Indians had embraced pigs in an attempt to preserve their own ways. In the 1650s an Indian sachem, accused of stealing an Englishman's hogs, countered with the charge that the colonists had killed the Indians' deer. The colonists told him that the hogs had been marked as private property by notches cut into their ears, while the deer had no similar marks. "Tis true indeed, none of my deer are marked," the chief responded, "and by that [you] may know them to be mine: and when you meet with any that are marked, you may do with them what you please; for they are none of mine." The story, perhaps apocryphal, points to a truth: North America had originally belonged to the Indians, but the colonists had claimed it as their own.

In the early years, coexistence seemed possible between colonists and Indians. Just as they negotiated the trade of animal skins and military alliances, colonists and Indians found ways to use the land together. Colonists helped Indians build fences around their fields and sometimes paid restitution for damage done by English livestock to Indian crops. Many laws mandated that free-ranging pigs be yoked—fitted with a large wooden collar to prevent their crawling under fences—or ringed, with a holly sprig or metal wire twisted through the nose to discourage rooting. Even these laws, however, were ignored or weakly enforced. Often pigs were simply pushed further away from colonial settlements; New Haven, for instance, exiled pigs to five miles outside town, where they could hurt only the crops of the Indians.

As the English population grew, colonists proved even less accommodating to their native neighbors. Rather than contenting themselves to live alongside Indians, the British tried to drive them away—and sometimes used pigs as a weapon in this effort. By 1663 Connecticut farmers were burning the fences around Indian cornfields so livestock could enter and destroy the crops. At about the same time, Maryland farmers earned the right, codified in a treaty, to shoot on sight any Indian caught stealing cows or pigs. Accusations that Indians had stolen livestock provided a pretext for attacks on native villages.

Animals proved capable of forcing Indians off the land all by themselves. Livestock served as the vanguard of empire: the free-range husbandry practiced by settlers expanded the colonial footprint because a constellation of hungry animals orbited around each settlement. Pigs ravaged Indian crops in the field. They dug up the baskets of grain Indians buried for future use. They trampled and ate reeds and grasses used for weaving. They ate the nuts and berries that Indians gathered for their own food. They devoured tuckahoe, a starchy root that Indians counted on when the corn crop failed. Along the coast, pigs despoiled oyster beds and clam banks. Roger Williams observed that pigs lingered near the ocean shore to "watch the low water (as the Indian women do)" and then rushed out onto the mud flats to "dig and root" for clams. Cotton Mather, addressing the theory that Indians might be descendants of the lost tribe of Israel, noted that Indians had "a great unkindness for our swine," the result perhaps of a dim cultural memory of the Levitical pork prohibition. A more likely reason Indians disliked swine, Mather admitted, was the animals' tendency to "devour the clams which are a dainty with them."

Native Americans initially had embraced the pig because it substituted for disappearing deer and bears and offered a chance at preserving some semblance of their way of life. At heart, though, the pig was their enemy: it helped destroy the landscape they relied on for sustenance.

Indians quickly came to understand this. "Our fathers had plenty of deer and skins, our plains were full of deer, as also our woods, and of turkeys, and our coves full of fish and fowl," a Narragansett sachem named Miantonomi explained in 1641. "But these English having gotten our land, they with scythes cut down the grass, and with axes fell the trees; their cows and horses eat the grass, and their hogs spoil our clam banks, and we shall all be starved." Mattagund, an Indian leader in Maryland, made a similar plea: "You come too near us to live & drive us from place to place," he wrote to the British in 1666. "We can fly no farther. Let us know where to live & how to be secured for the future from the hogs & cattle."

There was no place to flee. European settlers moved west with their livestock, driving native peoples from their land and claiming new territory for the American empire. Cows became the iconic animal of the American West, but in truth they arrived late to the scene. As the West was being won, it was pigs that gave pioneers the edge.

ELEVEN

"THE BENEVOLENT
TYRANNY OF THE PIG"

In March 1854 an Indian superintendent named Joel Palmer grew concerned about the Calapooya tribe living in the Willamette Valley of Oregon. The Indians' "principal means of subsistence," he wrote, was a type of marsh lily with "nutritious roots, once produced abundantly in the area." But European settlers had arrived with livestock, and because of the "increase in swine," which foraged for food in the marshes, the roots had disappeared. As a result, the Indians faced starvation.

And so it went from coast to coast. Britain had controlled all North American territory east of the Mississippi River since 1763, but for the rest of the colonial era, the Crown had banned settlers from crossing the Appalachian Mountains. After the Revolutionary War, citizens of the new United States celebrated their independence by striking out for the West. Over the next century, backwoods farmers cleared and settled far more land than the Germanic peoples of northern Europe had managed in a millennium. The American settlers did so using what geographer Terry Jordan-Bychkov has called "the four essential elements of backwoods farming": corn, axes, fire, and pigs.

In 1823 New England traveler Timothy Dwight defined pioneers as those who "begin the cultivation of the wilderness": they "cut down trees, build log-houses, lay open forested grounds to cultivation, and prepare the way for those who come after them," he explained. They practiced slash-and-burn agriculture, clearing fields with axes and fire and then planting corn amid the stumps. They ate corn as bread or drank it as moonshine. They hunted and trapped deer and smaller animals, fished in streams, and gathered wild nuts, berries, and greens. And they kept pigs by the dozen.

The settlers' stock-herding practices determined many of their other living arrangements. They lived not in villages but in homesteads scattered through the forest. One German observer noted in the 1780s that the likelihood of these settlers moving further west was "always increased by the preaching of the gospel." But in general they fled not so much from preachers as from people in general. To live off the land, they required a population density of two or fewer people per square mile. More than that cut into their hunting grounds and the rangeland for their pigs.

Early nineteenth-century travel diaries trace the spread of the pig across America. "Of all the domestic animals, hogs are the most numerous," François André Michaux reported from Kentucky in 1802. A traveler in Ohio in 1817 reported that pork could be had "in any quantity you please" because hogs "run in the woods in great droves." In Illinois a year later, an English visitor named Elias Pym Fordham commented on the fecundity of both humans and swine: "Every log cabin is swarming with half-naked children. Boys of 18 build huts, marry, and raise hogs and children at about the same expense." The children provided free farm labor; the pigs, free meat to sell at market. Fordham encouraged Englishmen to seek their fortunes in the New World: "If the industrious farmer invest his capital in land and hogs in Illinois, these will pay him 50 percent" annually as return on investment.

Pigs didn't linger much in the Great Plains—too few trees, too little water—but they had made it to the Pacific by the 1830s. By 1850 there were 30,000 in the Willamette Valley—about twenty-five per household—plus uncounted more roaming feral in the woods. During the California Gold Rush, these hogs were driven south to feed the miners.

The western poet Charles Badger Clark captured the importance of pork on the frontier in a bit of doggerel titled "Bacon":

> You're friendly to miner or puncher or priest;
> You're as good in December as May;
> You always came in when the fresh meat had ceased
> And the rough course of empire to westward was greased
> By the bacon we fried on the way.

A pioneer couldn't have asked for a better friend.

Travelers' accounts explain how those hogs were kept. In early Ohio, one man observed, "Hogs were almost as easily raised as the deer, and thousands were never seen by their owner until with his gun he went out and killed them." More often, farmers occasionally tossed out a little corn or salt to keep the pigs accustomed to being around people. Some pioneers, when they provided their animals with corn, also blew a horn or conch shell, thereby training the pigs, in Pavlovian style, to come running on command. In the spring, farmers roamed the woods to find new piglets and notch their ears as a mark of ownership. In the fall, one pioneer explained, he would build a pen and leave a gate open. "We put shelled corn in the pen and dribbled out a few long streaks through the woods," he explained. "Them half-wild hogs would foller the traces of corn up to the pen," and then he "would rush up and trap 'em." The pigs could then be fattened on surplus corn before slaughter.

Such swine exemplify an important fact about the species in general: although all pigs in the early United States were domestic, *Sus scrofa domesticus* was only a tiny bit removed from its wild ancestor, *Sus scrofa*. By raising their pigs in a semicontrolled manner, frontiersmen practically guaranteed that some of the animals would disappear into the woods to do what pigs do best: take care of themselves. Beginning in the colonial era, escapees like these turned feral: they reverted to their ancient ways and, within a few generations, lost the comparative docility of their domestic cousins.* In North Carolina, California, and elsewhere, the feral swine later interbred with pure-bred Eurasian wild boars that had escaped from exotic game parks. These swine, fecund as ever, created an enormous population of wild pigs that would haunt the backwoods—and eventually the suburbs—of the United States for decades to come.

---

*Some believe that America's feral swine are descended from the herd that accompanied Hernando De Soto's 1539 expedition. This is unlikely. Europeans who settled the South in the eighteenth and nineteenth centuries made no mention of encountering feral hogs. Had De Soto's hogs been breeding in the woods for hundreds of years, by 1800 they would have been more common than deer.

The semiwild forest pig lay at the center of American pioneer culture. Abraham Lincoln described himself as "a mast-fed lawyer," meaning that he picked up an education in backwoods districts, just as the local pigs fed themselves among the oaks and chestnuts. The pioneers had as many names for pigs as the Romans had for pork, most of them reflecting the animals' agility, toughness, and destructiveness. Woods pigs were called razorbacks, painters, rovers, thistle-diggers, prairie sharks, land sharks, land pikes, wind-splitters, hazel-splitters, sapling-splitters, rail-splitters, stump suckers, elm peelers, piney woods rooters, and—puzzlingly, but perhaps because they were so hard to get a grip on—cucumber seeds.

Mostly, though, pigs were called dinner. *Little House in the Big Woods*, Laura Ingalls Wilder's fictional account of her pioneer childhood, contains a loving description of roasting a pig tail after the fall slaughter. Pa skinned the tail and thrust a sharp stick into the wide end, Ma sprinkled it with salt, and the girls roasted it over hot coals. "Drops of fat dripped off it and blazed on the coals," Wilder writes. "It was nicely browned all over, and how good it smelled. They carried it into the yard to cool it, and even before it was cool enough they began tasting it and burned their tongues. They ate every little bit of meat off the bones."

Most reports of backwoods food were not nearly so complimentary. "In all my previous life I had never fallen in with any cooking so villainous," one traveler reported, describing meals of "rusty salt pork, boiled or fried . . . musty corn-meal dodgers . . . and sometimes what was most slanderously called coffee." Frederick Law Olmsted, a journalist before he became a landscape architect, referred to bacon and corn bread as "the bane of my life" during six months of travel in Texas. English geologist George William Featherstonhaugh, eating supper in

Arkansas, encountered "little pieces of pork swimming in hog's grease, some very badly made bread, and much worse coffee." Then he added a lament familiar to anyone who has traveled and eaten in remote places: "They knew very well that we had no other place to go to, and had prepared accordingly."

English writer Frances Trollope was more charitable. "The ordinary mode of living is abundant, if not delicate," she observed in 1832 after returning home from America. "They consume an extraordinary quantity of bacon." The archaeological record bears out those reports. When archaeologists dug up the bones from the Tennessee farmstead where Davy Crockett was born in 1786, more than 92 percent of the bones recovered came from pigs. Sites in the Ozarks dating to a few decades later show similar patterns.

Pioneers in the nineteenth-century United States relied heavily on pigs, but this was nothing new. A similar dynamic had also been at work at other times and places throughout history. In the Near East during the Iron Age, pig bones were exceedingly rare in nearly all communities. Then they make a sudden appearance in a few places around 1200 BC, precisely the time that the Philistines first settled the area. A mysterious group of "sea peoples," likely from the Aegean, the Philistines colonized new territory in Palestine and brought pigs with them. They raised swine during the early years of settlement, then later turned to other sorts of livestock better suited to arid conditions.

More than 1,000 years later, in the fifth century AD, the Anglo-Saxons conquered Britain and settled the countryside. At the village of West Stow in Suffolk, pigs account for a high percentage of the bones dating from the years immediately after settlement, even though the area, mostly grassland with few trees, was more appropriate for sheep and cattle. Only the pig could breed quickly enough to feed the new settlers. In later centuries, once the herds of sheep had built up, the number of pigs declined.

The evidence from the ancient Near East, Anglo-Saxon England, the colonial Americas, and the early United States all point to the same conclusion: the pig is the perfect animal for colonization, breeding quickly and providing abundant meat in the difficult years when the land is being tamed. One writer explained that in pioneer-era Minnesota, only when farms were well established could settlers start to raise cattle and "emancipate themselves from the benevolent tyranny of the pig." Cows and sheep are animals for more settled times. When the West was being won, America counted on the pig.

In most instances, the pioneer pig enjoyed only a brief moment in the sun. The Anglo-Saxon settlement at West Stow, after its initial pig-heavy period, turned to sheep. The Spanish in Latin America, once the native peoples had been conquered, began raising both cattle and sheep. Farmers in the eastern United States, likewise, made a switch to cattle after workers—free and slave—became available to tend the herds.

Midwestern states might have made a similar transition after the pioneer phase. But they didn't. The pig remained king for one reason: corn.

TWELVE

"TWENTY BUSHELS OF CORN ON FOUR LEGS"

Thaddeus Harris, a New Englander, traveled along the Ohio River around 1800 and noticed two types of settlers, one on each side of the river. "Here, in Ohio, they are intelligent, industrious, and thriving; there, on the back skirts of Virginia, ignorant, lazy, and poor," he wrote. "Here, the buildings are neat, though small, and furnished in many instances with brick chimneys and glass windows; there the habitations are miserable cabins." The Virginians' problem, Harris believed, was that they lived off the land: "The great abundance of wild game allures them to the forest; and from it they obtain the greater part of their miserable subsistence. In consequence of this, they neglect the cultivation of their lands."

These stereotypes contain a kernel of truth. Most back-woods farmers, preferring isolation, kept moving west or found isolated pockets in the hills where they could live untroubled by the likes of Thaddeus Harris. When these pioneers cleared out, another type of settler moved in. Unsatisfied with mere subsistence farming, this second wave had more ambitious plans. On the south side of the Ohio River, Thaddeus Harris saw the vestiges of a pioneer past. On the north side, in Ohio, he witnessed the birth of the Corn Belt.

The new farmers settled the best farmland, generally river-bottom lands that Native Americans had cultivated for centuries. They leapfrogged their way west, settling the valleys of the Scioto, Miami, Wabash, and other rivers. By the 1850s better plows made it possible to break up the prairie sod; wetlands were drained, more trees were cleared, and the Corn Belt—a continuous, five-hundred-mile region stretching from Ohio to Iowa—was formed. It became, and remains, the agricultural heartland of America and one of the most productive regions the world has ever known, thanks to rich soil and a remarkably bountiful grain.

Corn, paired with pigs, fueled the rapid settlement of the United States. An acre of corn produced three to six times as much grain as an acre of wheat. One sown seed of wheat might yield 50 at harvest; a single corn kernel produced 150 to 300. Only rice—a far more labor-intensive crop—produced at similar rates. One scholar estimates that if Americans had planted wheat instead of corn in their march across the country, it would have taken them an extra century to reach the Rockies.

The farmers grew breathtaking amounts of corn, but they didn't eat it. They preferred wheat bread. In their minds, corn wasn't food—it was feed. The practice of fattening livestock with grain dates back at least to ancient Mesopotamia, but North American farmers were the first to apply it on a vast scale. Whereas South America had become a major meat exporter by

The grain grown in America's Corn Belt became food not for people but for livestock. "The hog is regarded as the most compact form in which the Indian corn crop of the States can be transported to market," a British visitor said. (Courtesy Boston Public Library)

raising cattle on grassland, midwestern farmers turned grassland into cornfields and fed the corn to hogs and cattle. Compared to a field of grass, a field of corn produces far more calories and therefore far more livestock. Today the practice of feeding hogs on corn has become standard worldwide, with Brazil and China adopting it as the most efficient way to satisfy populations growing hungrier for meat. And it all started in the Corn Belt, when farmers figured out that an Old World animal and a New World crop made a perfect match.

In nineteenth-century America, corn was too difficult to transport to become a cash crop, so farmers turned it into value-added products that were easier to sell: pigs and whiskey. Historians have often described feeding livestock and making whiskey as the "solution" to the problem of marketing the Midwest's great corn crop. This suggests that farmers planted huge fields of corn, harvested it, and then sat around stroking their beards, trying to figure out what to do with all the stuff. In fact, the first Corn Belt farmers knew what they were doing right from the start: the lure of growing rich by fattening livestock on corn had drawn them west in the first place.

A family named Renick was among the first to perfect a new livestock system. Starting out along the south branch of the Potomac River in Virginia, the Renicks raised cattle and pigs and drove them to market in Baltimore and Philadelphia. By 1805 they had moved to Ohio's Scioto Valley, near Chillicothe, where the land had three feet of rich, black soil that sprouted bumper crops of corn. Cows and hogs ate the grain, then walked to market in Philadelphia and Baltimore. In 1819, two members of the Renick family traveled west to scout out new farmland, pausing on a bluff overlooking the bottomlands of the Mississippi River opposite St. Louis, where the great Native American civilization at Cahokia once flourished. The land, the Renicks wrote, "wants nothing but industry and art to afford some of the finest farms that any country can boast of." Within a decade or two, that land was thickly planted with corn.

One of the Renicks later described their system of fattening cows and pigs together in the same fields: "The cattle were not housed or sheltered, but simply fed twice a day in the open lots of eight or ten acres each, with unhusked corn with the fodder, and followed by hogs to clean up the waste and offal." There were no barns, which kept capital costs down, and the feeding method reduced labor costs as well: rather than harvesting ears of corn, they "shocked" the stalks—gathered them into large upright bundles—and left them in the fields. The cattle then ate the green roughage, or silage, as well as the corn kernels off the ear. After the cows had eaten, it was the pigs' turn: they ate the kernels that had escaped the attention of the cows, as well as many that had not: cow digestive systems, adapted to cellulose, were inefficient at processing grains, and the pigs enjoyed a great deal of corn that had already made one trip through a cow. As historian Allan Bogue has explained, "For cattle-feeders the margin of profit was often represented by the nutriment that his hogs gleaned from the droppings of steers."

Although hogs and cattle were fattened together, they took separate paths to the feedlot. With cows, the need for pasture placed the greatest demand on capital, but this was also the easiest part of the process to outsource. Ohio farmers traveled to buy lean cattle in the West, where young steers spent their first couple of years grazing on rangeland. As bones, organs, and other unprofitable parts of their bodies grew, the cattle ate grass, the food they had evolved to digest, which was cheap or free. Then, nearly grown, they were sold to feedlot operators who stuffed them with corn, so that this more expensive food went directly into producing meat and fat. Beef cattle were born in the Far West, eaten in the cities of the East Coast, and fattened on feedlots in between.

Pigs, with their shorter life cycle and omnivorous appetite, found free food closer to home. In spring and summer they gleaned fields, foraged for roots and nuts in scattered plots of woodland, and ate whey at dairies or spent grains at whiskey distilleries. In the fall, they gorged on corn before slaughter. Nearly all beef cattlemen kept hogs to clean up after the cattle, but other farmers stuck to hogs only. In the most basic form of fattening, pigs were simply turned loose into a field of standing corn, where they knocked over the stalks and harvested the ears themselves. This was known as "hogging down" a field.

In histories of midwestern agriculture, cows get most of the attention. A book on the early years of the Corn Belt observes, "We would have difficulty saying whether the Ohio Valley was more of a beef-cattle empire than a hog empire." That book, nonetheless, is titled *Cattle Kingdom in the Ohio Valley*. Farmers who fattened more hogs than cattle still referred to themselves as cattlemen. The ancient prejudice in favor of cows and against pigs persisted, and perhaps that shouldn't surprise us: when one creature eats the feces of another, it's not difficult to guess which will garner more respect.

Though less prestigious, hogs were more profitable. "Hogs don't always carry the prestige of cattle, but you can't live on prestige," one farmer explained. The hog earned the nickname "mortgage lifter" because it freed so many Corn Belt farms from debt. If farmers wanted to turn corn into meat and meat into money, pigs did the job two or three times more efficiently than cows.

"What is a hog, but fifteen or twenty bushels of corn on four legs?" a visitor to Chicago asked in 1867. Farmers hoped that figure would be closer to fifteen than twenty: the less corn they used to fatten a hog, the more money they made. The key measure was the feed-conversion rate: the number of pounds of corn required to produce one pound of pork. The tighter the ratio, the higher the profit. Feed conversion hadn't much mattered when hogs fed themselves in the woods, but a grain diet changed the equation. American farmers in 1820 faced the same situation their English counterparts had a century or two before:

woods and wastes had been plowed under to grow crops, and the age of the forest pig came to an end. British farmers fed their pigs the wastes from distilleries and dairies, as well as surplus peas and beans grown as part of rotation schemes. American farmers fattened hogs almost exclusively on corn. But American and English farmers had one thing in common: both needed a hog that fattened quickly while confined to a pen.

After 1700 English farmers started importing Chinese hogs and crossing them with local types to create new breeds that could convert feed efficiently while weathering chilly English winters. The pig's quick reproductive cycle made it easy to select for these desirable traits. A single sow could produce, in one year, five to ten female piglets, and each of those female piglets reached reproductive age in less than a year. By carefully choosing which animals to breed, a farmer could produce rapid changes in his herd, especially if he stuck to the common practice of inbreeding. In 1790 an English agriculture writer noted that one farmer had used a single boar to breed all of his sows, including the boar's "daughters, and his daughters' daughters," and that the herd had been "highly improved, by this incestuous intercourse." By 1850 English farmers had created most of the popular modern English breeds—the types now known as "heritage"—including Yorkshire, Berkshire, Hampshire, Leicester, Woburn, and Suffolk. A swine expert noted that the new types were "indebted to the Asiatic swine for their present compactness of form, the readiness with which they fatten on a small quantity of food, and their early maturity."

Those English breeds made their way to America, where farmers used them as breeding stock to create types suitable for New World conditions. George Washington acquired some pigs of the Woburn breed in the 1790s, and before long importers had brought in most of the improved British breeds, as well as pigs from South America, Italy, Spain, Russia, China, and Southeast Asia. All were put to use in shaping new American breeds such as the Chester White and Duroc Jersey. In the Corn Belt, the key hog was the Poland China, developed in Ohio's Miami Valley. Only the second half of the breed's name is accurate: it is thought to be a cross of local wood hogs with Chinese, Berkshire, and Irish Grazier pigs, with no Polish types entering the mix. Although American breeders followed the English lead, they soon far outstripped the motherland. In 1840 there were more than 26 million hogs in the United States, compared to about 2 million in Great Britain.

Though each of the modern breeds had distinctive qualities, they also had much in common: compared to woods hogs all had thinner bones, shorter legs, and smaller heads, and they turned grain into meat more efficiently. Feed conversion is a complex process, but one key is having lengthy intestines, the better to extract every bit of nutrition from food before it exits the body. In wild boars, the ratio of intestinal to body length was about 10:1. In the common woods hog, it was 13:1. In improved Corn Belt hogs like Berkshires and Poland Chinas, it was 18:1. Whereas woods hogs took two or three years to reach market size, the new types reached slaughter weight at eighteen months or less. American pigs became the envy of the world. The English prided themselves on their cattle and sheep, but they were forced to admit that America had created better pigs. "Nowhere in the world can such marvelous herds of swine be found as in the corn states of America," a British official conceded. "Here the pig is monarch of all he surveys."

As intensive agriculture moved west, the razorback era came to an end. You could trace the advance of the frontier by the appearance of the pigs: the rangy forest hog became scarce in Pennsylvania by 1800, in Ohio by 1820, and in Illinois by 1840, pushed out by fatter hybrids.

The new hogs that replaced the forest pigs were valued for their bulk—but they couldn't be too fat. A pure Chinese type thrived as a backyard pig, slaughtered in the same place he was raised, but that didn't suit Corn Belt conditions in the nineteenth century, when a more nimble pig was needed.

Before trucks, the only way for pigs to leave the farm—for a rail depot, or a river port, or a slaughterhouse—was by walking. An agricultural newspaper explained that hogs "should have a straight hind leg" in order to "travel the required distance." Often that distance was long, a hundred miles or more. The Poland China dominated the early Corn Belt because, even after growing hefty, it remained agile enough to make the journey.

Nearly all livestock went to market on foot: cattle, horses, mules, sheep, goats, turkeys, ducks, and geese. ("That was the prettiest drive of anything they drove," a Tennessee old-timer said of geese. "They'd just paddle along on them webbed feet.") Hogs, though, ruled the road. Americans raised more pigs than any other type of animal, so naturally swine crowded out other beasts on the turnpikes. The best estimates suggest that in antebellum America, five times as many hogs were driven as all other animals combined. In 1847 one tollgate in North Carolina recorded 692 sheep, 898 cattle, 1,317 horses, and 51,753 hogs.

As the United States grew, traveling hordes of pigs crisscrossed the country in all directions. The farmers who rushed to settle the West after the Revolutionary War soon returned east with pigs to sell. Around 1800 some of the very first Corn Belt hogs were driven from Ohio farms to Baltimore slaughterhouses. Other hogs walked from Kentucky to Virginia, from the Nashville basin to Alabama, and from southern Illinois

HOG DROVERS.

Every winter hundreds of thousands of American pigs went to market on foot, many of them crossing mountain ranges and walking hundreds of miles. Long-distance hog droving, nearly forgotten today, was at least as significant—and trickier to pull off—than the far more famous cattle drives.

to Chicago. By the 1840s and 1850s, a growing rail network mostly ended the era of long-distance droving, but the railroad builders were stymied by the Blue Ridge Mountains, which separated the hog-raising regions of Kentucky and Tennessee from the pork-eating slave South. A few farmers from Lexington, Kentucky, walked their hogs through the Cumberland Gap and all the way to Charleston, South Carolina, a distance of more than five hundred miles.

Pig drives followed fairly standard protocols. The drover, on horseback, rode at the front of a herd that might range from a few hundred to 1,000 or more hogs. Following behind, on foot, were his employees, called drivers, usually one for every one hundred pigs. The drivers shouted, "Soo-eey," "Su-boy," or "Ho-o-o-yuh"—this last, according to one witness, was pronounced thusly: "The first syllable is like a prolonged wail, while the last syllable is hurled out with a snap and a thud, much like the exclamation one might make if suddenly hit in the solar plexus."

This was not easy work. Whenever a roadside creek or pond appeared, the pigs flopped into the mud and commenced wallowing. The secret, one drover said, lay in not exerting too much control: "Never let a hog know he's being driven. Just let him take his way, and keep him going in the right direction." The start of the journey was especially difficult, for during that stage loud noises could send pigs stampeding back toward their home farms. One solution was to sew up their eyelids: temporarily blinded, the pigs clumped together and kept to the road by feel. At their destination, the stitch was clipped and their vision restored. (The young Abraham Lincoln, charged with driving a recalcitrant drove of hogs aboard a riverboat, pulled out a needle and thread and started sewing eyelids.) After a few days on the road, the hogs settled into a routine, and the biggest problem became beasts who couldn't keep up. Lame pigs were traded to innkeepers for room and board.

Cows, sheep, and goats have been driven great distances for millennia because they move well in herds and require only grass or other greenery along the way. Driving pigs on such long journeys has been rare historically because it is more difficult: the animals not only needed shade and tended to scatter but also required provisions en route. Roman swineherds were among the few to take on such challenges. To fulfill the pork dole in the Roman Empire, tens of thousands of hogs walked well over a hundred miles to Rome from the forested regions of Campania, Samnium, and Lucania. We don't know many details of their journey, but we do know that they lost weight.

Weight loss on the road was known as "drift," and in the United States an infrastructure grew up to ensure it didn't happen. Because pigs could walk about ten miles a day, inns—often known as wagon stands—sprang up at ten-mile intervals along the roads, offering drovers and their pigs food and a place to sleep. At the taverns, the hogs were herded into corrals and given corn, usually eight bushels per one hundred hogs. One traveler described watching a drove of 1,000 hogs devour their evening meal: "The music made by this large number of hogs, in eating corn on a frosty night, I will never forget."

Because droving was a decentralized trade, it's impossible to know its full scale. It is clear, however, that hog drives were at least as significant as the more celebrated cattle drives. The largest cattle drives, from Texas to Kansas, involved as many as 600,000 cattle a year, but they lasted just fifteen years or so. Hog droving, by comparison, involved hundreds of thousands of animals during peak years and on some routes lasted nearly a century. From Kentucky alone, as many as 100,000 hogs per year were driven east to Richmond, Baltimore, and Philadelphia. In 1855 more than 83,000 hogs walked along a little-known route through Mount Airy, North Carolina. And there were many other routes: through Asheville along the French Broad River, from the Nashville basin along the Natchez Trace into Alabama, through Knoxville into Georgia, and out of Kentucky through the Cumberland Gap or along the Kanawha River. The route through the Cumberland Gap in Kentucky, known as the Wilderness Road in the years immediately after Daniel Boone blazed it, later came to be called the "Kaintuck Hog Road" after its most frequent traveler.

Once the drovers had crossed the mountains, they fanned out to sell their hogs, either at individual plantations or at local slaughterhouses. Such small-scale pork-packing plants operated all over the country, buying hogs from farmers or drovers and wholesaling the pork to merchants. With time, however, most of these small plants disappeared, forced out of business by a new phenomenon: centralized pork-packing operations that operated on a vast scale and with astonishing efficiency. As slaughterhouses grew larger and more highly mechanized, pigs—the freewheeling, self-sufficient creatures that had helped conquer a continent—took their first steps into a rigid, industrialized future.

THIRTEEN

"THE REPUBLIC OF PORKDOM"

Frances Trollope, wife of a struggling English barrister, moved to America with her young children in 1827. They spent some time in a utopian community in Tennessee, which soon failed, and then moved to Cincinnati, where she undertook business ventures that fared no better. In 1831 she moved back to England, became a writer, and found success with her first book, *Domestic Manners of the Americans*. Americans, she thought, were overconfident and undereducated, overly fond of spitting, and hypocritical: "You will see them with one hand hoisting the cap of liberty, and with the other flogging their slaves."

Though she did find a few things to like in Cincinnati—the lovely hills, a twin-spired brick church, a school for girls— she never adjusted to the city's principal industry. "I am sure I should have liked Cincinnati much better if the people had not dealt so very largely in hogs," she wrote. As she was on a stroll one day, her feet "literally got entangled in pigs' tails and jawbones." Seeking an escape, she rented a cottage outside the city but soon noticed new buildings being constructed nearby. "'Tis to be a slaughter-house for hogs," a man told her. Since there were nice homes nearby, Trollope asked if a slaughter-house might be legally prohibited as a nuisance.

> "A what?"
> "A nuisance," I repeated, and explained what I meant.
> "No, no," was the reply, "that may do very well for your tyrannical country, where a rich man's nose is more thought of than a poor man's mouth; but hogs be profit-able produce here, and we be too free for such a law as that."

Trollope had the ill fortune to live in Cincinnati at the start of a slaughterhouse boom. Hogs were indeed "profitable produce" in the Corn Belt, so much so that pork packing became one of America's most innovative industries. To deal with the enor-mous volume of hogs, packers pioneered the factory assembly line that Henry Ford later adapted for automobiles. To squeeze every bit of value from pig carcasses, new industries sprang up to transform by-products into soap and chemicals. Above all, the industry produced a great deal of meat.

Pork had been a favored food for millennia in part because it cured well. For the same reason, it became the key meat of global trade in the years before artificial refrigeration: unlike fresh beef, salted pork could be stored at room temperature for months as it was transported thousands of miles. By producing so much meat, Corn Belt packers transformed diets in America and around the world.

The conversion of pigs to pork began each fall. Hogs, fattened on the latest corn crop, traveled by foot, barge, or railcar to a packinghouse, almost always located on a river. The packers bought and slaughtered the hogs, cured the meat in salt, then in the spring, when the rivers thawed, floated the pork downriver to market. The packers invested in pigs, salt, barrels, and labor in November and sold their product in March or April, which meant that they saw no return on investment for five or six months. There was good money to be made in pork, but the risks and capital requirements were considerable and the competition stiff. Dozens of midsized packers were scattered along the Ohio and Mississippi Rivers and their tributaries, each slaughtering about 8,000 hogs annually.

In the 1820s the pigs and the money started to concentrate in Cincinnati, which became known as "Porkopolis." The most important port on the Ohio River, the city lay south of Ohio's Miami Valley, the first great center of Corn Belt hog raising and within droving distance of the corn-growing regions of Kentucky and Indiana. As the biggest city in the Corn Belt, Cincinnati offered enough skilled workers to fill the seasonal jobs of killing, cutting, and packing the swine that flooded the city each fall.

Cincinnati's fortunes rose and fell alongside those of New Orleans, another great port city. Nearly all of the Corn Belt's output made its way down the Ohio and Mississippi Rivers to New Orleans; from there it was shipped to plantations in the lower South or loaded onto oceangoing ships bound for Europe or America's eastern seaboard. Just as importantly, the rivers allowed salt to be brought north by steamboat, supplementing the meager local supplies. With the transportation infrastructure in place, eastern merchants brought their money to Cincinnati, and pork packing exploded. Between 1832 and 1841—the decade just after Frances Trollope left town—Cincinnati packed about 500,000 hogs. During the 1850s, that figure rose to 2.3 million. By the 1870s, it had reached 6 million.

By then, however, Cincinnati had been displaced: Chicago had become the new "hog butcher for the world," slaughtering as many as 4 million hogs in a single season. Chicago benefited from the westward shift of the Corn Belt into Illinois and Iowa. Just as important were trains. By the early 1850s, good railroads linked the Corn Belt to eastern cities, and Chicago lay at the terminus of multiple rail lines. With the coming of the Civil War, Chicago's triumph was complete: the city offered reliable transportation by rail just as fighting shut down river traffic and as the US Army began buying salt pork in quantity.

Though Chicago dominated the US pork market, it was not the only game in town. Cincinnati remained important, and the packing industry grew in St. Louis, Kansas City, Milwaukee, and Omaha. The telegraph lines that ran alongside railroad tracks gave farmers the most valuable of business commodities: information. Now farmers could drive or ship their pigs to the packinghouse offering the highest price per pound.

In order to attract farmers with high prices while still turning a large profit, packinghouses strove for efficiency. In 1838, after a visit to Cincinnati, English writer Harriet Martineau observed, "The division of labour is brought to as much perfection in these slaughter-houses as in the pin-manufactories of Birmingham," a reference to a famous passage in *The Wealth of Nations* in which Adam Smith explains that productivity rose enormously when the making of the simple pin was "divided into eighteen distinct operations." Twenty years later Frederick Law Olmsted described "a sort of human chopping-machine" at a Cincinnati plant. "No iron cog-wheels could work with more regular motion," he wrote. "*Plump* falls the hog upon the table, *chop, chop; chop, chop; chop, chop*, fall the cleavers. All is over. But, before you can say so, *plump, chop, chop; chop, chop; chop, chop*, sounds again. There is no pause for admiration. . . . We took out our watches and counted thirty-five seconds, from the moment when one hog touched the table until the next occupied its place."

Martineau and Olmsted observed efficient division of labor, but the dead hogs were still being carried by hand to the chopping block. Soon that would change, as the process became automated.

The modern pork-packing line, fully developed by the 1860s, started with pigs being driven up a ramp—known as the "Bridge of Sighs" in Chicago—to the top floor of a two- or three-story plant. A worker attached a chain to a rear leg of each pig and hooked the other end of the chain to an overhead rail, which hoisted the animal, kicking and squealing, into the air. The "sticker" plunged a knife into the pig's neck, and blood poured out of the wound and through a latticed floor, to be collected in barrels below. Now dead, the pig was plunged into a tank twenty feet long and six feet wide, filled with water kept near boiling by a continuous flow of steam. Men standing alongside used short poles to keep the line of hogs bobbing along the trough, rolling them over to ensure an even scald that would loosen the hair.

At the end of the tank a rake-like device lifted the carcass and dumped it onto an inclined table with eight or nine men on each side. Their job was to scrape off the hog's hair and bristles: cattle were skinned, but pork was sold skin-on because the skin was tender enough to eat. The first two men took off only the bristles—valuable for use in brushes—along the spine. The carcass then slid or rolled down to the next men in line, who used short-handled hoes or sharp knives to scrape and shave the hair from every bit of the pig. (By the 1890s, most of these hair-removing jobs had disappeared, taken over by automated scrapers.)

At the end of the table, two men attached a stick called a gambrel that stretched the rear legs apart, then hooked the gambrel to an overhead track—known as "the railroad"—so that the pig swung free of the table, hanging from its splayed feet. Propelled by gravity or a steam-driven chain, the pig carcass started down the line. At the first station, two men sprayed it with cold water to wash away loose hair and filth; then the railroad made a series of stops at four foot intervals: the "dry shaver" removed stray hairs; the gut man slashed the belly open with a single stroke, allowing the intestines to pour out; the organ man removed the heart, liver, and other innards; finally, another man with a hose sprayed out the interior of the carcass. Each worker had just twelve seconds to perform his task. The hog then rolled off to the cooling room to await butchery the next day.

The technology of this line was not complex. The idea was what mattered, and it was inspired by the nature of packing hogs, a messy job not easily mechanized. Pigs have complex shapes, and each is slightly different, so the work of killing, bleeding, gutting, cleaning, and cutting required the practiced eye of a human worker. There were two ways to make it more efficient: reduce the effort expended by workers in hoisting slippery carcasses and shorten the interval between each operation. The pork industry found ingenious ways of doing both.

The overhead rail marked an epochal moment in the history of factory work. The Ford Motor Company holds the credit for inventing the modern assembly line to make the Model T in 1913. An assembly line entails a number of features, including the subdivision of labor (at least as old as Adam Smith's pin factory in the 1770s) and interchangeable parts (developed by clock and gun makers even earlier). Pigs, of course, are living creatures that (at least before the era of genetic modification) lack the sort of anatomical consistency that would allow for this sort of processing. But pork packers in the nineteenth century did make one key innovation that set the stage for Ford's assembly lines: they remained in place while the item of manufacture—in this case, a hog—came to them.

Assembling cars and disassembling pigs had much in common. In most factories the worker "spends more of his time walking about [looking] for materials and tools than he does in working," Henry Ford noted drily in his autobiography. "He gets small pay because pedestrianism is not a highly paid line." Ford explained that his company found a solution in what he called the "overhead trolley" used by pork packers to carry dead hogs to a succession of workstations. A moving line—of pigs or cars—not only cut down on heavy lifting but also set the pace of the work. Each slaughterhouse worker had just seconds to perform his task before another pig arrived. It was rough on the workers, but the rate of production soared. The genius of the packers' disassembly line and of the Model T assembly line, Ford explained, lay in their bringing "the work to the men instead of the men to the work."

The pig disassembly line became a standard stop for journalists and tourists visiting Chicago. "Great as this wonderful city is in everything," a British traveler said, "the first place among its strong points must be given to the celerity and comprehensiveness of the Chicago style of killing hogs."

The "comprehensiveness" of Chicago's pork producers extended to their use of the carcass. In Upton Sinclair's *The Jungle*, a slaughterhouse employee says, "They use everything

Workers in Chicago chain live hogs to a wheel that
hoists them into the air for slaughter, just one part
of a sophisticated system invented to process the
enormous hog crop of the Corn Belt. Henry Ford
said that the idea of the assembly line was inspired
by a visit to a slaughterhouse. (Courtesy Library of
Congress)

about the hog except the squeal," and the novel's narrator
mocks him for using a stale "witticism." Though the expression
had already become a cliché by 1906, it was nonetheless true.
Snout-to-tail eating may have been the frugal tradition of peasants, but no one could squeeze all the value from a pig like a
profit-hungry pork packer.

Packers preferred pigs to be about two hundred pounds
and fattened on corn. Acorn-fed pigs, though prized today, were
considered second-rate in nineteenth-century America. Corn
produced firm, shelf-stable pork fat. An acorn diet, heavy in
unsaturated fats, made for soft, oily pork fat, which improves

the flavor of dry-cured hams but makes barreled pork—the most marketable type at the time—mushy and prone to rancidity. In 1837 corn-fed hogs sold for $5 per hundred pounds of live weight, while mast-fed brought only $3.

Most Corn Belt meat was placed in barrels and covered with a brine of salt, sugar, and saltpeter (potassium nitrate, a preservative that also imparted a pinkish color to otherwise grey meat). Known as "barreled pork," "pickled pork," or simply "pork," it appealed to buyers because of its cheapness and long shelf life. Pickled pork was divided into three classes. The highest quality, "clear pork," was sold mostly in New England, where cod and mackerel fishermen demanded the best as their shipboard provisions. The military and other large institutions primarily purchased "mess," the second-best quality. Slaves in the United States and the Caribbean ate the lowest quality, known as "prime."

Some pork—mainly bellies, whole sides, and shoulders— was not wet-packed in brine but rather dry-salted and smoked. This was called bacon. Though bacon in the United States today is brined pork belly, 150 years ago the word referred to dry-salted smoked pork, regardless of what part of the pig it came from. A visitor to the Illinois prairie in 1837, for instance, noted that the pioneers "make bacon of hams, shoulders, and middlings." In Cincinnati and Chicago, salt-rubbed pork was allowed to cure for a few weeks or months, cold-smoked over hickory, beech, and maple, then packed into gigantic boxes known as hogsheads, each containing eight or nine hundred pounds of meat. This bacon had a wide distribution all over the country and in Europe.

There was pork to suit every budget. Hams—dry-cured in salt and sugar or wet-cured in a brine of molasses, saltpeter, and salt—commanded the highest end of the pork market. At the lower end were feet and tongues, sold soaking in a spicy pickle. Pig heads were cooked down into headcheese, a jellied loaf with bits of meat. Organs, as well as meat from ribs and necks, were chopped, mixed with fat and spices, and injected into cleaned intestines to make sausage, which "enters largely into the subsistence of the laboring classes of society," one nineteenth-century observer wrote.

Fat was nearly as valuable as meat. Lard served as the primary cooking fat in the United States until the middle of the twentieth century and was exported in bulk to Latin America and Europe. The highest quality, leaf lard, came from around the organs and was used for baking. Lesser varieties were turned into industrial oils and grease. Some lard was separated into two parts, lard oil and stearin. The oil was used in lamps—it competed with whale oil in the pre-kerosene era—while stearin was turned into candles and soaps. Pork fat and its derivatives also became oleomargarine, carbon paper, roofing pitch, and explosives.

The list of by-products was nearly endless. Bristles became brushes, while finer hairs stuffed mattresses. "Tankage"—the solid bits strained out from rendered fats—was ground into feed for pigs and chickens. The contents of intestines became fertilizer. Bones were stamped into buttons, cooked to make gelatin, or smoldered into charcoal for use in refining sugar. Hooves were boiled down for glue. Prussian blue—a dye used by printers—was derived from blood, as was albumen for the photographic industry. Extracts from glands and organs—pepsin and other enzymes, various hormones, and more—served as raw materials for the pharmaceutical industry.

Pig by-products gave packers an incentive to grow large. Small-scale slaughtering didn't produce a marketable amount of blood, hair, or bones, but killing thousands of animals a day changed the equation. A large packinghouse might pay $10 for a live hog and sell its meat for $9.75, but it didn't lose money. The profit margin came from selling the by-products that small packers threw away. That meant big packers could pay higher prices for pigs, allowing them to force out smaller competitors. Cincinnati packers in the 1850s paid, on average, 5 percent more than competitors elsewhere. Farmers benefited. So did consumers, whose prodigious pork consumption was subsidized by the sale of by-products.

Since the colonial period, Americans had been famous for consuming vast amounts of beef and pork, especially by comparison with the meat-starved peasantry of Europe. Statistics for early America are hard to come by, but we have some good clues. In many wills, husbands specified the amount of meat their widows were to be given. Widows typically received 120 pounds annually in 1700; a century later, that figure had risen to over 200 pounds. In the antebellum South, a typical ration for a slave was 3 pounds of pork per week, or about 150 pounds per year. Laborers in the North ate 170 pounds or more. After 1900, the statistics become more reliable. Between 1900 and 1909, per capita meat consumption in America was about 170 pounds, compared to 120 pounds in Britain, 105 in Germany, and 81 in France. "There are a great many ill conveniences here, but no empty bellies," one Irishman in America wrote to his family back home.

The United States in 1900 saw itself as a nation of beef eaters, but that reflected aspiration more than reality: not until the 1950s did per capita beef consumption surpass that of pork. Pork was the meat of rural dwellers and the poor, while urbanites and the more affluent ate beef. This difference had to do with population density and technology. Beef was best eaten fresh, not salted, and artificial refrigeration at home was uncommon until after World War I. For a butcher to sell fresh meat from a nine-hundred-pound steer, he needed the large customer base that only an urban area could provide—and even in 1900 only two out of five Americans were city dwellers. Most lived in the country and stored their own meat supplies at room temperature. That meant salt pork.

Americans in the nineteenth century got most of their meat and fat from pigs. In his 1845 novel *The Chainbearer*, James Fennimore Cooper notes that a family is "in a desperate way when the mother can see the bottom of the pork-barrel." (This sentiment underlies our expression "scraping the bottom of the barrel.") Pickled pork lurked in nearly every dish. In Eliza Leslie's *Directions for Cookery*, one of the most popular cookbooks of the nineteenth century, the recipe for pork and beans—"a homely dish, but . . . much liked"—called for a quart of beans and two pounds of salt pork, and her chowder contained as much pork as fish. One man, recalling his midwestern childhood, described a typical rural diet: "For breakfast we had bacon, ham, or sausage; for dinner smoked or pickled pork; for supper ham, sausage, headcheese, or some other kind of pork delicacy." As a physician wrote in the magazine *Godey's Lady's Book* in 1860, "The United States of America might properly be called the great Hog-eating Confederacy, or the Republic of Porkdom."

Cheap American pork helped change the menu in Europe as well, as an important part of a growing international food trade that greatly improved the diet of Europe's peasants and industrial workers. After enjoying a spell of relative prosperity following the Black Death, the European peasantry by 1500 had returned to a scanty grain-based diet, punctuated at frequent intervals by famine. Hunger and starvation were common through 1800, but by the nineteenth century the situation had improved. European farmers adopted intensive crop rotations and better technology. The belated embrace of potatoes and corn—two high-yielding New World crops—boosted the number of calories available. Thanks to better technology, canned vegetables, meat, and milk became safer, cheaper, and at least tolerably palatable. Most importantly, improved trade allowed the mass importation of grain and meat from the Americas and the Antipodes.

Meat production flourished wherever cheap feed could be had. The United States, with both rangeland and corn, specialized in beef and pork. Argentina's grasslands made it a leading producer of beef and mutton. Australia and New Zealand, similarly blessed with rangeland, exported primarily mutton. The beef and mutton trade got a boost in the late nineteenth century with the rise of artificial refrigeration, which allowed chilled and frozen meat to move around the world. Pork too was shipped fresh, though more of it was cured.

The British benefited most from this global trade. Whereas most European countries placed high tariffs on meat imports to protect local farmers, Britain kept her ports open and reaped the benefits of cheap meat. The upper class bought refrigerated beef from Argentina, the middle class bought frozen mutton from Australia, and the poor made do with cheap bacon from the United States—but, one way or another, nearly all Britons had meat on their plates.

As shipping technology improved and global trade expanded, imports brought down the cost of food and improved the nutrition of the working classes of the Western world. Workers who once had spent 50 to 75 percent of their income on food now spent 25 percent or less. And they were eating better.

In the West, diseases caused by vitamin and mineral deficiencies became less common. The average height of adults—an indicator of nutrition levels in childhood—increased by several inches, making the Englishman of 1910 seem like a giant compared to his countryman from a century before.

In the midst of this commercialization, however, some people preserved the old ways. The pig had proven itself fit for industrial production and global trade, but it didn't lose its place as the favored livestock for those who raised their own meat. English cottagers kept pigs in backyard sties. In the American South, the landless poor ran their hogs in the woods. And even in the heart of Victorian cities, pigs scavenged the streets and wound up on the dinner tables of the poor.

FOURTEEN

"A SWINISH MULTITUDE"

One April morning in 1825, two hog catchers went to work in New York City's Eighth Ward, in what is now Greenwich Village. They were accompanied by four city marshals because they were expecting trouble—and they found it.

As they rounded up stray pigs and locked them in their cart, the hog catchers attracted what a newspaper described as "a large mob of disorderly people" who demanded the return of their livestock. Someone threw a brick that hit one of the marshals in the face, and the crowd rushed the cart and "let loose all of the hogs, who quickly scampered off."

This was just one skirmish in Manhattan's hog wars. On one side were poor city dwellers—mostly English, Irish, and African American—who raised pigs for food. The city's streets functioned as an urban commons that provided food for "the defenseless poor," one advocate explained. On the other side was a rising middle class who saw the animals as a public nui-sance—dangerous to children—and upsetting to ladies who might glimpse swine copulating in the street. The odds of that happening were good: in 1820 some 20,000 hogs lived in Man-hattan, about one pig for every five people. The pigs devoured "all kinds of refuse," a Norwegian visitor noted. "And then, when these walking sewers are properly filled up they are butch-ered and provide a real treat for the dinner-table."

Dining upon a "walking sewer" struck the visitor as foul, but the owners of city pigs could not afford to be fastidious. If they wanted meat, they had to raise their own. This remained true all over the United States and Europe throughout the nine-teenth century and into the twentieth. Pigs scavenged on the streets of New York and London, wandered free in the piney woods of Georgia and North Carolina, and wallowed in the sties of small farmers from Maine to Lincolnshire. Like the ur-ban poor in Mesopotamia and Egypt, these poor Americans and Englishmen kept pigs as a buffer against hardship but came under increasing threat from more powerful people who, for various reasons, wanted to take their animals away.

There have been city pigs as long as there have been cities, and the United States carried on the Old World practice. Colonial New England towns appointed "hog reeves" to corral unrestrained beasts, and a Massachusetts court in 1658 warned, "Many children are exposed to great dangers of loss of life or limb though the ravenousness of swine." Though such fears were not without basis—consider the killer pigs of medieval Europe—the laws were generally ignored. As in earlier eras, pigs provided not only food but also the only functioning urban sanitation service.

"Once more in Broadway," Charles Dickens wrote of his 1842 visit to New York. "Take care of the pigs. Two portly sows are trotting up behind this carriage," searching out a meal of "cabbage-stalks and offal." Dickens described the pigs as "ugly brutes" with "scanty, brown" hair and "long, gaunt, legs." These were not chubby Corn Belt porkers but rangy scavenger types, fit for doing battle with stray dogs and mean boys, just as forest pigs could fight off the wolves that had hunted them in early America and medieval Europe.

Loose pigs in the streets, according to another English visitor to New York, "would arouse the indignation of any but Americans." In fact, English cities hosted plenty of pigs, and many people there were indeed indignant about them. In *The Condition of the Working-Class in England*, Friedrich Engels called attention to "the multitude of pigs walking about in all the alleys, rooting in the offal heaps, or kept imprisoned in small pens" in the working-class districts of Manchester. Living amid filth, though, was the price of having meat on the table. In the Potteries, a large slum within the wealthy London suburb of North Kensington, an 1851 census turned up three pigs for every one person.

The upper classes conflated the poor with their pigs. In *Reflections on the Revolution in France*, Edmund Burke wrote that if democracy prevailed, "learning will be cast into the mire, and trodden down under the hoofs of a swinish multitude." The phrase quickly became one of the era's most popular epithets, a way of vilifying the lower orders by equating them with the most abject of animals. As the wealthy saw it, both the poor and their pigs bred quickly, lived in filth, and threatened the social order.

The authorities could do little about the presence of the swinish multitude, whose members were needed to work the mills

Pigs roamed the streets of many nineteenth-century cities, playing a dual role as sanitation service and food for the poor—while also offending delicate middle-class sensibilities. Health concerns finally forced authorities to banish pigs, as shown here in an 1859 illustration from a New York newspaper.

and mines, but they could banish the swine. In the 1860s and 1870s, public health measures forced most pigs from England's large cities.

By that time most pigs had disappeared from New York as well. In 1849 a cholera epidemic prompted New York to take sanitation seriously. Hogs hadn't caused cholera—a water supply contaminated by human sewage did—but they were swept up in the general cleaning frenzy. City officials, who in earlier decades had backed down from outraged pig keepers, stood their ground this time. New York's professional police force, only recently created, led an effort that drove off more than 5,000 swine. Manhattan, at least in its more densely developed southern stretches, became pig-free. The poor had to find other ways to feed themselves.

The common folk of the American South, unlike their counterparts in the industrializing North, managed to hold onto their pigs until well after the Civil War. This improved the diets of poor southerners but did not help their reputations. When travelers ventured into the South, they viewed poor country folk just as they viewed poor city folk: with contempt. Southerners "delight in their present low, lazy, sluttish, heathenish, hellish life, and seem not desirous of changing it," one English visitor wrote. He was not speaking about the owners of plantations, who had grown rich on the labor of enslaved human beings. He was describing, rather, the plain white folk of the South, who lived humbly but ate well.

Like the rest of the country, southern states preserved an open range during the colonial period, granting livestock free run of all unfenced land. Most states in the East and Midwest changed course by 1850 or so, giving deed holders full control over their land. The South, though, preserved its free-range customs until after the Civil War and in some areas until after World War II. Even the politically powerful railroads had to pay damages when their trains killed livestock. When a railroad company argued before the Georgia Supreme Court in 1860 that a plaintiff had been negligent in allowing his horse to wander into the path of an oncoming train, the justices rejected the argument. If this were true, the court wrote, then a "man could not walk across his neighbor's unenclosed land; nor allow his horse or his hog or his cow to range in the woods." The era of the "Private Property: No Trespassing" sign was still far in the future.

In the South the common people's "rights in the woods"— which allowed them to hunt, fish, and keep livestock on any unfenced land—were held sacred, because without those rights many would have starved. Despite the presence of cotton, rice, and sugar plantations, less than 15 percent of southern land had been cleared in 1850. That left millions of acres available for hunting and herding. The 1850 census revealed that, per capita, there were more swine in the South than in the Corn Belt, suggesting that nearly every southerner acquired a few pigs, notched their ears—the equivalent of branding—and turned them loose in the forest. And the census surely undercounted southern pigs, since owners could easily lower their tax bills by underreporting their free-ranging beasts. "You can keep as many pigs as you wish, and you need not feed them," a German visitor wrote in a letter home. "We can live here like lords." A prosperous farmer in the Blue Ridge Mountains of North Carolina described how impoverished men could move to the region and, with the aid of "free mast for their hogs," build "a nice little happy home." He concluded, "Truly, it is a paradise for the poor man."

The open range shaped the cuisine of the South. Pork and corn were staples on the East Coast during colonial times and in the West in the early nineteenth century, but these other regions diversified their diets after the pioneer phase passed. The South did not, instead preserving frontier eating habits into modern times. Large planters imported pork and cornmeal from the Corn Belt, while smaller farmers persisted in their old ways, surviving on the produce of their gardens, their pigs, and wild game. On special occasions, both rich and poor dug a pit, built a fire, and cooked a whole hog, low and slow, over the embers. Barbecue—both the word and the technique were adopted from the Taino Indians of the Caribbean—became popular in Virginia by the 1750s and spread west with the settlers. The barbecue became the standard southern outdoor celebration for occasions such as political rallies or July Fourth celebrations. As one English visitor to America explained, "A barbecued hog in the woods, and plenty of whiskey, will buy birthrights and secure elections." Cattle and sheep were barbecued on occasion, but pigs proved most popular—they tasted best, and there were plenty of them ranging in the woods.

Southern leaders began to close the range after the Civil War, and their main purpose was to take food out of the mouths of the poor. The abolition of slavery had created a labor shortage in the southern plantation economy. The planters needed field workers, and most freed slaves had no desire to work for their former masters. Under prevailing law, freed African Americans might have adopted the habits of plain white folk and earned an independent living through herding, hunting, and fishing. A newspaper editor in Virginia explained the implications of this for the southern economy. Laborers, he wrote, if "furnished with free food, would neglect agriculture." That prospect frightened the white elite.

Legislatures in every state closed the range on a county-by-county basis, starting in areas with the largest populations of freed slaves. A newspaper claimed that this change affected only "lousy negroes and lazy white men" and would be good for both: now they would find work as sharecroppers and tenant farmers. This was indeed the effect, although few, black or white, would find reason to celebrate it: like the enclosure movement in England, the closing of the range stripped the poor of a means of self-sufficiency and left them vulnerable to exploitation by wealthy planters.

As poor southerners were forced into the Reconstruction economy, pig populations began to dwindle. Historians who examined six counties in Alabama and Mississippi found that, before the Civil War, residents had owned 2.1 hogs per capita. After the war, it was 0.4. The effect on the common folk was plain to see. If travelers to the South before the Civil War tended to mock southerners for their laziness, those who came after the war pitied these same people for their poverty, misery, and hunger. In the 1880s, one writer described the closing of the range as "another step in the oppression of the poor."

The banning of urban pigs and the closing of the southern range hurt the landless poor, but many people still had access to a little property. They couldn't raise cattle or sheep, which required pasture, but they could keep a pig in a sty. In nineteenth-century England, perhaps a quarter to half of rural workers did so. "Life without a pig was almost unthinkable," a Buckinghamshire man observed. "To have a sty in the garden . . . was held to be as essential to the happiness of a newly married couple as a living room or a bedroom." Living so close to the home, the pig became a sort of edible pet, a source of companionship as well as food. The pig was "one of the best friends of the poor," according to an English authority in 1806.

Small-scale pig keeping followed age-old rhythms. Cottagers generally bought a spring piglet after it had been weaned, then fed it until early the next winter. Children were sent to gather wild plants in the spring and summer, and in the fall they collected acorns and beechnuts. Kitchen scraps were tossed into a wooden "pig tub" by the back door, and potatoes filled out the provisions. In their final few weeks, the pigs ate barley to harden the fat before slaughter.

When raised close to home, swine sometime became objects of affection. "The pig was an important member of the family," an Oxfordshire woman reported, "and its health and condition were regularly reported in letters to children away from home, together with news of their brothers and sisters." One cottager kept a careful account of his expenditures on a pig he sent to market and, upon selling it, calculated that he had made three

A Virginia farmer and his hogs in 1939. Hog slaughter functioned as a crucial seasonal ritual for centuries in Europe and America. Fattened on the bounty of summer and fall, pigs provided a crucial source of meat and fat during the lean times of winter. (Marion Post Wolcott; courtesy of the Library of Congress)

shillings. "Not much profit there," he was told. "No," the man replied. "But there: I had his company fer six months!"

The same fondness for swine could be found among wealthy landowners who raised enormously fat pigs for competition at agricultural fairs. No one enjoyed a pig's company more than Lord Emsworth, hero of a series of comic novels by P. G. Wodehouse and owner of a prize-winning Berkshire sow named the Empress of Blandings. "Watching her now as she tucked into a sort of hash of bran, acorns, potatoes, linseed, and swill," Wodehouse writes, "the ninth Earl of Emsworth felt his heart leap up in much the same way as that of the poet Wordsworth used to do when he beheld a rainbow in the sky."

E. B. White, who lived on a small farm in Maine, captured this affection in a more heartfelt way in an essay titled "The

Death of a Pig." Buying a spring pig, feeding it through the
summer and fall, and slaughtering it in early winter "is a trag-
edy enacted on most farms with perfect fidelity to the original script," White writes. "The murder, being premeditated, is
in the first degree but is quick and skillful, and the smoked
bacon and ham provide a ceremonial ending whose fitness is
seldom questioned." One year, however, White's pig fell sick.
He sought veterinary advice and administered castor oil and
later an enema—"Once having given a pig an enema there is no
turning back, no chance of resuming one of life's more stereo-
typed roles"—all to no avail. The pig died, White grieved, and
his neighbors shared his sorrow. "The premature expiration of
a pig is, I soon discovered, a departure which the community
marks solemnly on its calendar, a sorrow in which it feels fully
involved," he wrote. "The loss we felt was not the loss of ham
but the loss of pig. He had evidently become precious to me, not
that he represented a distant nourishment in a hungry time, but
that he had suffered in a suffering world."

White, a man of means, had no need to worry about the
loss of ham. For poorer men, the death of a pig was a financial
as well as an emotional blow. According to an English observer,
"A man had virtually one chance only of ever adding to his cash
income, and that was by raising more than one pig." A study
of Oxfordshire suggested that raising pigs constituted 12 to 15
percent of a laborer's income. "Pig clubs," a sort of mutual insurance program, collected dues from members and paid out
if a pig died of illness. In *Middlemarch*, George Eliot defines a
happy village as one where "nobody's pig had died."

Those pigs that survived illness were slaughtered right
where they had been raised, a process that unnerved people
more often as pigs became ever rarer in domestic life. In England
most cottagers hired skilled pig killers, who would bring along

knives and a scalding tub. The pig was lashed to a bench or held tight with a noose around its snout. Although sometimes stunned with a hammer, more often it was simply stuck in the neck and allowed to bleed out, with the blood caught in a pan and cooked into black pudding.

Memoirs of rural life in England and America often describe such events. "The killing of the pig was the great event in the domestic life of the year," one man remembered. Neighbors helped neighbors and shared in the bounty and fun. In *Little House in the Big Woods*, Pa removes the bladder from a freshly slaughtered pig, blows it up like a balloon, and hands it to his daughters, who joyfully bat it back and forth before returning to the labor of rendering lard and making sausage.* Pig killing was a communal ritual, a break in the rhythms of daily life, a sign of the passing seasons. It was a solemn occasion—the pig was a friend and did not want to die—and a time of celebration. This drama was peculiar to pigs because cattle and sheep were rarely kept around the house. Only pigs were coddled and then killed, their horrifying, humanlike shrieks piercing the neighborhood. One girl recalled that, during the slaughter, she would "creep back into bed and cry," remembering how she had fed cabbage stalks to her beloved swine. The next day, however, she happily dipped her bread into pork gravy made from that same pig's flesh. She was just a girl, she said, "learning to live in this world of compromises."

Thomas Hardy devotes an entire chapter of *Jude the Obscure* to the killing of a pig. Jude and his ill-matched wife, Arabella, have hired a pig killer who fails to appear, so they undertake the task themselves, lashing the pig to a stool with its legs in the air.

---

* Inflated swine bladders were used as balls in many sports, which explains how the American football earned the slightly inaccurate name "pigskin."

"Upon my soul I would sooner have gone without the pig than have had this to do!" said Jude. "A creature I have fed with my own hands."

"Don't be such a tender-hearted fool! There's the sticking-knife—the one with the point. Now whatever you do, don't stick un too deep."

"I'll stick un effectually, so as to make short work of it. . . ."

"You must not!" she cried. "The meat must be well bled, and to do that he must die slow. We shall lose a shilling a score if the meat is red and bloody! Just touch the vein, that's all. I was brought up to it, and I know. Every good butcher keeps un bleeding long. He ought to be eight or ten minutes dying, at least."

"He shall not be half a minute if I can help it, however the meat may look," said Jude, determinedly. Scraping the bristles from the pig's upturned throat, as he had seen the butchers do, he slit the fat; then plunged in the knife with all his might.

"'Od damn it all!" she cried, "that ever I should say it! You've over-stuck un! And I telling you all the time—"

"Do be quiet, Arabella, and have a little pity on the creature!"

. . .

The dying animal's cry assumed its third and final tone, the shriek of agony; his glazing eyes riveting themselves on Arabella with the eloquently keen reproach of a creature recognizing at last the treachery of those who had seemed his only friends.

"It's a hateful business," Jude says, but Arabella replies, "Pigs must be killed. . . . Poor folks must live."

In 1895, when *Jude the Obscure* was published, few would have disputed Arabella's view. For all but a tiny fringe of the population, the slaughter of livestock counted not as cruelty but as necessity. Jude's horror was a distinctly modern reaction, one that soon would grow more common. As people moved to cities and bought their meat at stores, home slaughter became rare and upsetting. Rather than forming part of the rhythm of home life, the act of killing animals now took place far away, in slaughterhouses. That distance, though, carried a cost. The walls of slaughterhouses hid not only the act of killing but also a multitude of other sins.

FIFTEEN

"A GROWING PREJUDICE AGAINST PORK"

*/*The ear was assailed by a most terrifying shriek: the visitors started in alarm, the women turned pale and shrank back." So begins a famous passage in Upton Sinclair's 1906 novel *The Jungle*. The visitors have chosen to take a tour of a slaughterhouse. They watch as a worker hooks chains to the legs of pigs and an overhead rail lifts them into the air: "Another was swung up, and then another, and another, until there was a double line of them, each dangling by a foot and kicking in frenzy." The narrator is impressed by the efficiency, but also appalled. "It was pork-making by machinery, pork-making by applied mathematics," he explains, then continues: "And yet somehow the most matter-of-fact person could not help thinking of the hogs; they were so innocent, they came so very trustingly; and they were so very human in their protests—and so perfectly within their

rights! . . . One could not stand and watch very long without becoming philosophical, without beginning to deal in symbols and similes, and to hear the hog-squeal of the universe."

Most Americans, in fact, could hear such squeals without becoming philosophical. A newspaper described Sinclair's concern for pig suffering as "nauseous hogwash," and the author himself later disavowed any such empathy, claiming he had intended the passage as "hilarious farce."

Sinclair often struggled to convey his intended messages. A failed author of romantic novels and a recent convert to socialism, he had traveled to Chicago to examine working conditions at slaughterhouses. He hoped that his novel, serialized in a socialist journal in 1905 and published as a book a year later, would expose the plight of exploited workers and prompt a revolution. He failed in that goal. "I aimed at the public's heart," he explained, "and by accident I hit it in the stomach."

The American people, Sinclair was not the first or last to learn, had a large capacity for ignoring the sufferings of the less fortunate. The food on their plate, however, was a different matter, especially after *The Jungle* offered this description of sausage making:

> There was never the least attention paid to what was cut up for sausage; there would come all the way back from Europe old sausage that had been rejected, and that was moldy and white—it would be dosed with borax and glycerine, and dumped into the hoppers, and made over again for home consumption. There would be meat that had tumbled out on the floor, in the dirt and sawdust, where the workers had tramped and spit uncounted billions of consumption germs. There would be meat stored in great piles in rooms; and the water from leaky roofs would drip over it, and thousands of

rats would race about on it. It was too dark in these storage places to see well, but a man could run his hand over these piles of meat and sweep off handfuls of the dried dung of rats. These rats were nuisances, and the packers would put poisoned bread out for them; they would die, and then rats, bread, and meat would go into the hoppers together.

From the hoppers emerged sausage that was "sent out to the public's breakfast."

Though *The Jungle* spelled trouble for the meat industry in general, it was especially bad news for pork. Long regarded as the food of poor people and country folk, it was now increasingly rejected by a nation growing wealthier and more urban. A newer concern was trichinosis, the recently identified parasite, transmitted through undercooked pork, that encysted its larvae in human muscles. That seemed like a big risk for the sake of a pork chop.

The meat industry fought hard to counteract such prejudices: it cleaned up its plants, the government helped craft new rules to combat trichinosis, created new ways to cure and market ham and bacon, and even enlisted the wives of farmers to promote pork in newspaper columns and at grocery stores. Such efforts had mixed results at best.

Even before *The Jungle*, the public had grown suspicious of the Chicago meatpackers, who had consolidated into a cartel that fixed prices, set production levels, and divided territory, controlling most of the industry from slaughter to retail. After a rise in retail meat prices—partly beyond the meatpackers' control—infuriated the public, the federal government in 1902

After the scandals triggered by *The Jungle*, meatpackers tried to convince the public that their products were sanitary and wholesome—as in this 1912 advertisement showing a fresh-faced blonde girl in white clothes, with ham and bacon carefully wrapped in white paper and proudly displaying the government inspection seal.

successfully sued the meat trust under the Sherman Antitrust Act. The victory proved only nominal: the big companies absorbed the costs, skirted the restrictions, and continued to operate as before. *The Jungle* altered the debate. Previously, consumers had worried about being ripped off. Now they feared being poisoned.

Newspapers and magazines jumped on the muckraking bandwagon and found more evidence of tainted food. Meat-packers had been using borax—a mineral most often found in detergents—to preserve meats, which explained how one firm could advertise an additive that would maintain the freshness of "pork and liver sausage, when exposed on your counter, and in the hottest weather, for at least one week." Newspaper reports of such practices fueled public outrage. President Theodore Roosevelt, ill-disposed to trusts in general and the meat industry in particular, commissioned an investigation that confirmed most of *The Jungle*'s allegations. Within four months of the book's publication, Congress passed the Pure Food and Drug Act and the Meat Inspection Act. The federal government took over inspection of all packing plants, promising that sanitation standards would be enforced and that no "unsound, unwholesome" meat would reach the American public.

As it turned out, less meat of any sort found its way onto Americans' plates. In the decade after *The Jungle*'s publication, per capita meat consumption plunged from 170 to 140 pounds a year, and it would remain relatively low for decades. Two world wars and the Great Depression played a role in that decline, as did the growing number of urbanites who rejected breakfast meats in favor of cereals like Kellogg's Corn Flakes. Meatpackers, though, blamed *The Jungle* and what one industry official called "systematic anti-meat propaganda."

Beef eventually rebounded from this dip in popularity, but pork continued to struggle. Its associations were largely negative and had deep cultural roots. According to Edward Hitchcock, a chemist and president of Amherst College, bacon "is so extremely undigestible and heavy" that it should be eaten only by the "laboring classes" and shunned by "the sedentary and the literary." "Fat bacon and pork are peculiarly appropriate for negroes," a physician explained in 1860. The medical theories of the sixteenth century—that only manual laborers could properly digest pork—were alive and well in the nineteenth. This was thanks in no small part to Sylvester Graham, popular health reformer and father of the eponymous cracker, who embraced Renaissance medical writers in his campaign to persuade Americans to eat less meat.

Simple dislike buttressed these medical theories. America had inherited from England a hierarchy of meats that placed beef and veal at the top, lamb and mutton next, and pork at the bottom. One cookbook writer dismissed barreled pork as "sea junk"—a reference to its use as a maritime provision—and rejected its taste as "villainous," while another described pork as "dangerously unwholesome." An 1893 guide to household management claimed, "A growing prejudice against pork in all its varieties . . . pervades our best classes."

Statistics bear out these observations. A 1909 study of 8,000 families in US cities found that wealth shaped the type of meat people ate. The highest income group ate three times as much poultry and 50 percent more beef, compared to the lowest income group. The poor ate the most pork. African Americans ate more pork than whites, and as their income rose, they spent the extra money not on more pork but on beef and chicken. For southern whites, the same was true: as they began to earn more money, their pork consumption stayed flat as their consumption of chicken and beef climbed.

Pork packers, well aware of these trends, responded with new marketing tactics. Rather than selling anonymous barrels of pickled pork, they followed the cultural trend toward national brands marketed directly to consumers. This was one area where pork had an advantage. Beef, because it cured poorly and had to be sold fresh in the butcher case, could not easily embrace this model. Pork products—salted, smoked, and wrapped in branded packaging—could. Customers at butcher shops requested porterhouse steaks or hamburger—company of origin unknown—but they learned to ask by name for Armour's Star Bacon and Swift's Premium Ham.

In their efforts to improve the reputation of pork, meat packers emphasized ham and bacon. The focus on ham was not surprising because it had always been the most prestigious cut. American hams—especially Virginia's Smithfield variety, from peanut-fed hogs—earned so much praise that Queen Victoria had a standing order for six Smithfield hams a week. Those were dry-cured or country hams, and packers had once produced them in quantity even though the cure took months. In the twentieth century, however, they switched most production to the city ham, wet cured in a brine solution much like barreled pork. To speed the process—lengthy cures tied up a lot of capital—packers started injecting brine into the ham with needles. Later they invented "vein-pumping," which involved blasting brine from a high-pressure hose into a large vein in the ham and allowing the animal's circulatory system to spread it through the meat. Such methods cut a three-month cure down to a week, then later to just a few hours. Efficiency triumphed, but flavor suffered. Dry-cured hams achieve greatness over a period of months as enzymes break down proteins into dozens of intense flavor compounds. Wet-cured hams tend to be soggy and insipid.

Bacon received an even more thorough reinvention. In 1850 the term "bacon" applied to any dry-cured cut of pork; fifty years later, as packers standardized their terminology, "bacon" referred only to belly meat—and nearly all of it was wet cured. The earlier practice of salt-packing pork bellies for more than a month, then smoking them, required too much time for packers operating on an industrial scale. A sweet pickle delivered better results: bellies were dumped into 1,000-gallon vats holding a sugary brine, then drained and moved to the smokehouse. The method could not deliver the intense flavor of dry-cured bacon, but it did reduce labor costs, produce a more consistent product, and give Americans the sweetness they craved even in meat.

Before World War I, this new type of bacon was cut into slabs of at least four pounds, wrapped in waxed paper, and branded with the company label. Customers sliced it at home. But in 1915 some producers began to sell wet-cured bacon presliced in one-pound packages. By the 1950s, packers had automated their lines so that bellies were pressed to uniform thickness, needle-injected with brine, automatically sliced, shuffled into the familiar shingle-like display, and packed in clear plastic that let customers view the streaks of lean and fat. The automated processes invented to produce modern bacon required heavy capital investment, but they paid off. By 1960 bacon had shed its reputation as a country meat and been reborn as a beloved breakfast staple for all classes.

Fresh pork experienced no similar resurgence. It didn't help that consumers were instructed to cook it until well-done—which generally meant bone-dry—in order to kill disease-causing worms. The lifecycle of *Trichinella spiralis* starts when a host—a person or pig, for instance—eats meat that contains the worm's encysted larvae. The host's gut digests the cyst walls, releasing the larvae, which grow into sexually mature adults, mate, and produce more larvae, which enter the bloodstream and then the muscles, where they encyst themselves and wait for another creature to eat them. Cooking meat to 137 degrees Fahrenheit kills the larvae, but not every cook followed that rule, leading to headlines such as "Missouri Town Reports 47 Cases of Trichinosis."

Pigs generally got the worms the same way humans did: by eating undercooked pork. Up through the 1950s and 1960s, feeding garbage to pigs on a commercial scale was common. (When in-sink garbage grinders such as the DisposAll became popular in the 1960s, they earned the nickname "mechanical pigs" because they got rid of food waste, a job recently held by genuine pigs.) Garbage-feeding swine clustered around the highest concentrations of garbage, in cities on the East and West Coasts. *The American City*, a policy journal, surveyed garbage-collection methods in 1920 and found that nearly a third of cities with populations over 100,000 used swine feeding as their primary garbage-disposal method; an even higher percentage of smaller cities did so. The article pointed to the economic benefits: "A ration of 1 pound of marketable pork to 50 pounds of garbage has been established, and with pork at 20 cents a pound on the hoof . . . garbage as feed is worth $8 a ton." The article cited, as an example, Worcester, Massachusetts, which had a population of 185,000 and kept a herd of 2,000 pigs to process its waste. In a period of just over two years, the city made a profit of $59,000 selling garbage-fed pork.

The large-scale practice of feeding pigs garbage and then selling the meat became more common during World War II, when corn became expensive and growers sought other sources of feed. Secaucus, New Jersey, home to 250,000 hogs, earned the nickname "Pig Capital of the East." The farmers collected garbage from Manhattan restaurants each night, fed it to their pigs, and sold the pork back to Manhattan restaurants. The farms survived until 1960, then fell victim to encroachment by the New Jersey Turnpike and neighbors' complaints about the stench.

It was a model of efficient waste disposal—except that the garbage often contained pork scraps. Pigs, of course, ate meat, and swine cannibalism on a large scale had been common in

America for more than a century because packers sold pork by-products as swine feed. Those by-products, however, had been cooked and dried, rendering them disease-free. Restaurant garbage, by contrast, might contain raw pork, which could transmit not only trichinosis but devastating livestock diseases like hoof-and-mouth and African swine fever. "Human trichinosis is based almost entirely on porcine trichinosis," a 1942 New York State commission concluded. "And porcine trichinosis is based almost entirely upon feeding hogs raw garbage containing trichinae-infested pork scraps." In 1952 a national outbreak of a swine disease called vesicular exanthema was traced back to garbage-fed pigs, and the US Department of Agriculture (USDA) took action. A 1952 rule required that all garbage fed to pigs must first be cooked to kill pathogens.

Garbage-fed pigs constituted a tiny portion of the annual pork crop—less than 2 percent—but they caused an outsized portion of the problems. "Although garbage-fed hogs are daily sold as food universally, there is some aversion to this practice," one report noted mildly. Meat quality suffered: because pigs lay down fat in much the same form in which they consume it, the taste of their flesh is closely linked to diet. And because the supplies of garbage- and corn-fed hogs were not differentiated in the market, a bad pork chop reflected on the entire industry. Lingering fears of parasites proved even more damaging. As one scientific study noted, "The prevalence of trichinosis in the United States has long cast an unfavorable light on the production of American pork."

Despite the best efforts of America's growing pork industry, the dubious practices of some pig farmers and meatpackers prevented the reputation of pork from rebounding. A 1942 study from the USDA noted "a shift in the consumption of meats as incomes rose, from pork to beef, veal, and lamb." In other words, despite the remaking of ham and bacon, pork remained the meat of the poor. A 1955 study of urban consumers found that with each step up in income level, beef consumption rose while pork consumption fell. Distinctions between city and country dwellers persisted as well. Urbanites devoted half of their meat consumption to beef and only a quarter to pork, while in the country those percentage were reversed.

These trends did not bode well for the pork industry. World War II factory jobs accelerated the population shift from country to city, and city dwellers made more money than rural Americans. A booming post–World War II economy raised living standards. After the war, three out of four American families had mechanical refrigerators, which meant they could now keep fresh meat in their homes and had less need to store cured meat in the pantry. Beef, most palatable when fresh, could suddenly be enjoyed with much greater convenience, a shift that undermined one of the greatest reasons for pork's historic appeal.

The year 1953 marked the end of an era for the American pig. That year, for the first time, Americans ate more beef than pork: 77.6 pounds of beef per capita, compared to 63.5 pounds of pork. The trend would continue in the decades ahead. By the 1970s, pork consumption had fallen to fifty-one pounds, while beef rose to eighty-six pounds.

The pork industry took note. The president of the National Pork Producers Council traveled the country to meet with pig farmers and give a talk titled "Improving the Image of Pork and Pork Producers." In the 1960s the Porkettes, a women's auxiliary to the Iowa Swine Producers Association, assigned themselves a similar task. They created a mascot, Lady Loinette, who served as a partner to the primary pork mascot, Sir Hamalot.

The Iowa Pork Queen, crowned each year by a committee of Porkettes, served as an ambassador for the meat. One queen asked, "Who first but Iowa would envision combining the image of the hog with the enthusiasm of vibrant young women in order to promote the pork industry?" The Porkettes held contests for baking with lard and created a campaign to promote pig leather with the slogan "Pigskin—Our Prettiest Byproduct" (which may have served as a reminder of how distinctly unpretty the other by-products were). The group's magazine, *Ladies Pork Journal*, included a feature titled "You'll Know She's a Porkette When . . . ," with such responses as "She passes out new pork recipes to 'city' friends."

Americans' pork prejudices persisted nonetheless. The first president of the Porkettes told a story about a "professional man from the city" who visited her farmhouse and told her he did not eat pork because it was "unwholesome." Another Porkette was conducting a grocery store promotion when an "old grandmotherly lady" explained that she always served applesauce as a side dish "to counteract the poison in pork."

The American pork industry was floundering. It would take all the brainpower and ingenuity of the American government, universities, advertising firms, and pharmaceutical companies to set it right. By the 1960s, those resources were at the ready. Over the next four decades, the pork industry changed what pigs ate, where they lived, and how fat they grew. While they were at it, the experts went ahead and changed the color of pork from red to white.

SIXTEEN

"THE OTHER
WHITE MEAT"

In 1986 leaders of the National Pork Producers Council gathered in a darkened room to hear their advertising agency pitch a new industry tagline: "Pork—the Other White Meat." When the lights came on after the two-hour presentation, the pork producers found themselves "in a state of shock," one executive recalled. Hog farmers, along with everybody else, had always viewed pork as a red meat, in competition with beef. Now they were being asked to spend good money promoting it as an alternative to chicken. According to *National Hog Farmer*, many thought it was a "dumb idea."

But these were desperate times, so pork producers took the plunge. Since the 1970s sales of poultry had soared as consumption of beef and pork plunged. Studies linking red meat to heart disease and cancer had taken a toll, and Americans had become fearful of fat. In one survey more than a third of Americans agreed with the statement "Pork would be a good meat except for the fat." The new campaign would convince people that pork was not bloody and fatty like beef but pale and lean like chicken.

With ice-skating star Peggy Fleming as spokeswoman, the pork industry launched the new marketing campaign at a January 1987 New York press event attended by the editor of *Better Homes & Gardens* and national television news reporters. Before the year was out, the advertising bill ran to more than $9 million. Almost immediately, the campaign was deemed a success. Eight out of ten Americans recognized the phrase "the other white meat," which lodged itself in that special place in the American mind that holds slogans like "Got Milk?" and "Just Do It." In 2011 *Adweek* deemed the campaign "among the most successful rebranding moves in the history of the food biz."

But it was more than a rebranding. The new slogan marked the culmination of a transformation in American farming. In 1945 pigs, bred by small farmers and raised outdoors on corn, grew thick layers of fat under their skin. By 1985, raised indoors on scientifically formulated feed and bred to exacting standards by large corporations, they produced very lean meat. The same qualities that suited pigs to small-scale production—fecundity and rapid growth—also made them perfect for industrial farming. In seeking to rebrand their product, pork producers had not just changed their tagline. They had created a new pig.

The Corn Belt was home to the "lard-type" hog, as opposed to the "bacon type" or "meat type." The leaner meat hogs—which included breeds like the Danish Landrace, Tamworth, and Large Yorkshire—had a thin layer of back fat and were often cured as bacon for the British market. The primary producers of these bacon-type pigs were Denmark, Canada, and Ireland, where pigs ate protein-rich dairy by-products that promoted lean muscle growth. Pigs that ate mostly corn—higher in carbohydrates than protein—ran to fat, which is why the Corn Belt became the center of global lard production.

Corn Belt farmers historically had depended on the "lard-type" breeds—Poland China, Berkshire, Chester White, and Duroc Jersey—in response to market demands. Bulk purchasers of barreled meat, which was used to feed miners, sailors, and slaves, preferred fatty meat because it preserved better. There was also a big demand for lard as an industrial lubricant and cooking fat. Under some market conditions, a pig's fat was more valuable than its flesh, and packers dumped whole hogs into the rendering vats, wasting all of the meat in order to extract the precious fat.

Lard, however, became increasingly less valuable, a shift that started in the late nineteenth century and accelerated with each passing decade. After John D. Rockefeller's Standard Oil Company developed the oil fields of Pennsylvania, factory workers began to oil their machines with petroleum products rather than animal fats. Thanks to better technology for both canned food and artificial refrigeration, sailors and laborers could enjoy foods other than fatty pork. More people turned to vegetable oils such as soybean, peanut, and corn, which allowed a simple production cycle—grow plants and extract their oil—rather than the extra step required with animals: grow plants, feed plants to pigs, extract fat from pigs. Health concerns about animal fats arose after World War II, and brands such as Crisco advertised their vegetable shortenings as healthier than animal fat. All of these factors contributed to a single result: demand for lard plummeted, and so did its price.

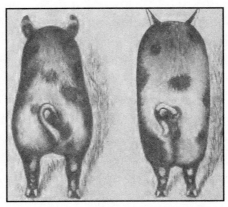

As consumers turned away from lard and fatty meats in the years after World War II, the government and meatpackers encouraged farmers to raise leaner pigs like the one at left in this diagram from a 1971 USDA pamphlet. Meat-type hogs were the first step on the road to "the other white meat." (Courtesy US Department of Agriculture)

In response, scientists and farmers worked to breed leaner hogs. Their model was Denmark, the first specialist in intensive hog production and America's key rival in the global pork market. In 1907 the Danes had created swine testing stations to carefully monitor feed intake and carcass quality, allowing them to choose breeding stock from those animals that gained the most lean muscle while eating the least feed. American agricultural colleges developed similar testing programs, and the USDA created new genetic lines to distribute to farmers. The meatpacker Hormel awarded prizes to farmers who raised the leanest pigs, and the private breed registries changed their standards as well. The Hampshire registry, for instance, specified that hogs should have no more than 1.8 inches of back fat and a pork chop measuring at least four inches square. In the 1950s, a 180-pound hog carcass yielded 35 pounds of lard. By the 1970s, a pig of the same size produced just 20 pounds of lard.

As America's pigs changed, so did the farms where they were raised. Before World War II American pigs lived almost precisely as they had a century before. They roamed on pasture in the spring and summer, grazing on crops such as clover or alfalfa. Farmers supplemented their diet with whatever was cheapest—one feed manual mentioned wheat middlings, bran, molasses, beet pulp, brewers' grains, sorghum, potatoes, cassava, rapeseed, whey, blood meal, and dozens of other foods. In the fall, hogs were fattened on corn. Nearly all farmers practiced mixed farming: they raised dairy cows, beef cattle, and hogs and grew crops to feed to their livestock. Diversification acted as insurance: if hogs were selling low, beef or dairy might be high. This small-scale system was enormously productive. In 1938 the United States raised 62 million hogs, compared to 39 million for all of western Europe.

Soon those pigs would be eating a different sort of diet, one that paired corn with soybeans—while also adding standard doses of antibiotics. During the meat shortages of World War II, scientists at the University of California researched a fact long known in Asian cultures: soybeans provided a concentrated source of high-quality protein. Paired with carbohydrate-rich corn, soy became an excellent animal feed. But it wasn't perfect. Researchers found that pigs gained weight most efficiently only if their corn-soy feed was supplemented with an animal-derived protein such as skim milk, fish meal, or slaughterhouse by-products. Since the 1920s scientists had understood the dietary importance

of vitamins A, D, and E. They hoped to discover a similar nutri-
ent, provisionally named "Animal Protein Factor," to explain the
growth patterns of pigs. In 1948, researchers at pharmaceutical
firm Merck and Company announced that they had isolated
the mysterious agent: it was a new vitamin, B_{12}. By 1949 feed
companies and agricultural colleges were promoting B_{12} as a
feed additive that could replace animal protein supplements.

Pharmaceutical companies needed a cheap source of B_{12},
and they found one close at hand. The microorganisms used
to grow streptomycin and other antibiotics also generated B_{12},
which remained as a by-product in production vats after the
desired antibiotic agents had been skimmed off. This B_{12} became a supplement in hog feed. Farmers soon developed a
strong preference for this particular B_{12} supplement. Animals
fed B_{12} that had originated as an antibiotic by-product gained
weight much faster than those given B_{12} from other sources.
Tests soon revealed why: B_{12} sold as pure had in fact contained
antibiotics. The big boost in growth rate came not from B_{12}
but from the drugs.

Soon antibiotics became the supplement of choice for American farmers. Further research showed that pigs given low doses
of antibiotics gained as much as 13 percent more weight than
pigs given the same amount of feed without the drugs. Pfizer
claimed that pigs dosed with Terramycin reached market weight
seventeen days faster and with far less feed, saving farmers a lot
of money. With little examination, the Food and Drug Administration in the 1950s approved the use of several antibiotics as a
feed supplement. By the 1960s, livestock consumed 1.2 million
pounds of the drugs each year. By the late 1990s, that figure had
risen to 25 million pounds.

Using antibiotics to promote growth had a side effect popular with farmers: the drugs helped ward off illness. This fact

became more significant as the next major shift in hog farming took place. At about the same time that antibiotics emerged as a feed supplement, farmers started pulling their hogs off pasture and crowding them together in barns. Land prices began to rise quickly in the mid-1950s, which made pasture too precious for pigs. With farmland so expensive, one farmer asked, "How can you afford to let pigs run around on it?" It made more sense to plow under the hog pasture, plant corn and soy, and feed those crops to pigs, now confined to barns.

The disadvantage of this plan was disease, since hogs kept in close quarters passed around illness more easily. That's where the antibiotic supplements came in. "The situation is sort of like kids at school," a pharmaceutical representative explained. "You know, one kid gets sick with the sniffles, and then all of 'em get it." Antibiotics, the drug salesman said, help pigs "start healthy, stay healthy, and gain as much weight as they can." In 1972 an agricultural magazine predicted, "The modern hog business would collapse without antibiotics."

Antibiotics helped address the problem of illness in confined pigs, but another trouble remained: labor costs. Pigs on pasture harvested their own food and deposited their manure broadly across the ground. Once pigs were confined to a barn, workers had to deliver food and remove waste. Farmhands, though, were scarce in the 1950s and 1960s, when a humming national economy attracted rural labor to the city. Farmers, who were smart businessmen, knew what to do. When labor is scarce, substitute capital. If you can't hire men to scrape stalls and haul feed, then buy machines to do it. Turn your farm into a factory.

In the shift to mechanized production, pigs followed the lead of poultry. As a seasonal, highly perishable crop raised outdoors on a small scale, chickens had historically been expensive. Birds raised indoors suffered from "leg weakness," later identified as rickets. Scientists learned, however, that adding vitamin D to feed solved this problem, and by the 1930s chickens were being raised indoors year-round. As electrical lines stretched into rural areas, farmers bought electric brooders, automatic feeders, and other labor-saving devices. Like pigs, chickens ate a corn-soy blend laced with antibiotics to speed growth. Scientists at agricultural colleges got to work on breeding, producing varieties that tolerated indoor life and quickly gained weight. Compared to its ancestor in the 1930s, a broiler chicken in the 1990s grew to twice the size in less than half the time. Once reserved for special occasions, chicken became affordable enough for everyday eating. Thanks to low prices and a healthy image, poultry consumption tripled between the 1940s and the 1990s, then kept growing.

Pigs and chickens have much in common: they eat similar diets (chickens, like pigs, are omnivorous) and grow to slaughter weight quickly—in less than two months for chickens and less than six months for pigs. Unlike cattle, which require many leisurely months on pasture, chickens and pigs can be stuffed with feed and turned into meat in short order. Unsurprisingly, then, the agricultural methods developed for one also worked well for the other.

In the 1950s and 1960s, American farmers started raising hogs like chickens. Feed, augered into the hog barns from nearby silos, was deposited in automatic feeders. Heaters and fans controlled the temperature, eliminating the need to open and close windows or haul straw for bedding. The most important innovation was low-tech: slatted floors. Used first in Norway in 1951 and adopted in the United States a decade later, the floors had long, narrow gaps that allowed urine and manure to fall into gutters below, where it could be flushed out with water. "The use of slatted floors has probably accelerated the trend toward confinement more than any other single development," an expert wrote in 1972.

Slatted floors started a cascade of other changes in pig husbandry. Straw bedding, formerly needed to absorb urine and provide warmth, could be eliminated in favor of bare floors. There was no need for a separate dunging area, so more pigs could be packed into pens where they slept, ate, and relieved themselves. For each pig weighing 150 to 250 pounds, industry guidelines in the 1980s called for allotting eight square feet of pen space, a dramatic reduction for animals that had historically been given free range of the woods or, at least, a pasture or sty. In such close quarters, pigs kept each other warm, requiring less artificial heat, and gained weight more quickly because they didn't burn calories exercising. Crowded together, they shuffled around more, trampling manure through the slots and keeping the pen cleaner.

Slatted floors made the farmer's life easier, eliminating what one industry publication called the "tedious and disagreeable" task of scraping manure from stalls. Once a solid that needed to be shoveled, manure became a liquid that could be sluiced away. The "comfort and convenience" of the farmer, an industrial manual reported, "may well be the most important" reason to move hogs into confinement. The comfort and convenience of the pigs was left unmentioned.

By the 1980s the most advanced farms practiced "life-cycle housing," which meant pigs never felt mud beneath their hooves. In a celebratory cover story in *Scientific American*, a university expert explained that the modern pig "lives indoors for its entire brief life: born and suckled in a farrowing unit and raised to slaughter weight in a nursery and later in a growing-feeding unit. It is fed a computer-formulated diet based on cornmeal and soybean meal with supplements of protein, minerals and vitamins. . . . It is sent to market at five or six months of age, having reached the slaughter weight of 220 pounds or more from its birth weight of two pounds."

The hog confinement building had become standardized by the 1990s. Roughly three hundred feet long, sixty feet wide, and made of metal, it sat on concrete foundations. The slatted floors, usually made of cast concrete but occasionally of metal or plastic, were kept bare, with no straw or other bedding. Waste fell to gutters below that drained into adjacent ponds, known as "manure lagoons." The buildings had no windows, instead relying on ventilation systems that operated automatically in response to temperature and humidity levels. Lights were kept off to save on electricity, except on the rare occasions when a worker was in the barn. Video systems allowed remote monitoring of the facilities.

Breeding became tightly controlled. By 2000, three-quarters of sows in the United States became pregnant through artificial insemination. The companies that sold boar semen—and the university researchers who collaborated with them—relied on advanced population genetics and performance testing of offspring. Breeding stock tended to be purebreds—primarily Berkshire, Chester White, Duroc, Hampshire, Poland China, Spot, and Yorkshire—that were interbred to create fast-growing hybrids.

Sows, rather than being kept together in group pens, spent their lives in "gestation crates," metal pens about seven feet long and two feet wide. This, the industry said, prevented them from fighting each other and kept dominant sows from monopolizing the food supply. When the four-month gestation period was nearly over, the sows were moved to "farrowing crates," about the same size but with a "creep rail": the lowest bar on one or both sides of the crate was missing so that piglets could enter the mother's pen to nurse but could also escape into a narrow area alongside that the sow couldn't enter. This alleviated a problem called "overlaying," in which the sow accidentally crushed her piglets. In a farrowing crate, the mother could stand up and lie down, but she couldn't turn around or roll over, lessening the chances of crushing. "If a sow has a litter of twelve and rolls on three, right there you've lost about a hundred dollars," one farmer explained in the 1980s. The piglets were weaned after two to four weeks, and the sow returned to the gestation barn to receive a fresh tube of semen. Some sows produced five litters in two years.

Most piglets had the tips of their daggerlike "needle teeth," or incisors, clipped to prevent injuries to the sow or to other piglets. They got an injection of iron and had their ears notched for identification. Their tails were docked because confinement hogs had a tendency to chew the tails of other pigs, perhaps out of boredom or anxiety. Male pigs were castrated during the first couple of weeks of life to prevent "boar taint," an off-flavor, caused in part by a pheromone, that sometimes appears in the meat of male pigs that reach sexual maturity before slaughter.

After weaning, the piglets started eating a carefully formulated feed. In the 1930s, pigs gained a pound of weight for every four pounds of feed they ate. In the 1980s, that pound of gain required three and a half pounds of feed. Today, it takes less than three pounds. As one animal scientist explained, "The wonders of genetic selection, improvements in animal health products, a better understanding of nutrition, and use of environmentally controlled barns has allowed animal scientists, veterinarians, and engineers to create these improvements."

Scientists and farmers had created "the other white meat." Although the phrase originated in a marketing slogan, in one sense it was literally true. Scientists from Texas A&M University showed that a factory-farmed pork loin's levels of myoglobin—the protein responsible for redness—were much lower than those of beef and comparable to those in chicken or fish. That pale color owed much to the new farming methods. Myoglobin carries oxygen to working muscles: the less a muscle works, the paler its color. Chicken breasts are white because those muscles are never used for flying. Crated veal is pale because the calf had no room to walk. Pigs raised on pasture had meat of a darker hue. By the 1980s pork had become a white meat because confinement pigs, packed into small pens, rarely used their muscles.

Whether customers preferred meat from these new pigs was a different question. After an initial boost spurred by the campaign, pork sales leveled off. People started thinking of pork as white meat, but they didn't start buying much more of it: per capita pork consumption stayed level at just under fifty pounds from the 1980s through the 2000s. Meanwhile, consumption of chicken—the original white meat—kept climbing, from fifty-four to sixty-nine pounds per capita.

In all of this innovation, one aspect of pork production was ignored: flavor. At an industry conference in the 1960s, an animal scientist at Oklahoma State University observed that in the rush to create lean pigs, "pork quality has been completely ignored by swine breeders." In 2000, industry experts writing in *National Hog Farmer* came to the same conclusion: "Currently, industry breeding schemes create pigs that grow fast and efficiently but lack the superior meat and eating quality consumers prefer."

One quality problem, identified in the 1960s and still unsolved, is "pale, soft, and exudative" pork, which is gray, mushy, and tasteless. This meat, it turned out, came from skinny, neurotic pigs. "Their personalities are completely different," the animal scientist Temple Grandin wrote of the lean pigs. "They're super-nervous and high-strung," and the stress appeared to be damaging their meat. Such pigs also had a tendency to drop dead of shock. As a group of veterinarians explained, such pigs "show an increase in carcass lean but much greater susceptibility to sudden death."

In creating lean pigs, American pork producers had created a new set of problems, of which meat quality was only the most obvious. Modern pig farming, many critics charged, destroyed small farms, fouled land and water, and threatened public health. Most of all, the critics said, the new farms made pigs miserable—a charge that, by the turn of the twenty-first century, was becoming harder to refute.

SEVENTEEN

VICES

While examining a group of hogs at a confinement facility around 1990, a veterinarian noticed a pregnant sow that appeared to be in pain. He asked about her and learned that she had broken her leg the day before and that her piglets were due in a week. "We'll let her farrow in here," the plant manager said, "and then we'll shoot her and foster off her pigs." The vet offered to splint her leg for free, but the manager turned him down. With 5,000 sows and just three full-time employees, he didn't have the manpower to care for a sow with a leg splint.

The vet was appalled. He had grown up on a hog farm. His father would have either treated the sow or euthanized her immediately, not allowed her to suffer for a week and then give birth. "If it is not feasible to do this in a confinement operation," the vet concluded, then "there is something wrong with confinement operations."

As the twentieth century wound to a close, many people were coming to that same conclusion. The *New York Times* printed its first analysis of modern pig farming in 1980, head-lined "Hog Production Swept by Agricultural Revolution," which pointed to the many smaller farmers being forced out of the business. A few years later writer Orville Schell published *Modern Meat*, raising concerns that feeding antibiotics to pigs might create drug-resistant pathogens. In 1996 the *Raleigh News and Observer* earned a Pulitzer Prize for its five-part series investigating how North Carolina's state government, captive to hog barons, had allowed the industry to pollute with impunity.

Then, in the late 1990s, the public learned about "mad cow disease." The infection, new to science, was contracted by cows that ate parts of other cows in their feed, and it could be passed along to humans who ate infected beef. The horror of a brain-destroying disease, caught from fast-food hamburgers, called attention to the appalling details of livestock production. Critiques of factory farms became a staple of investigative jour-nalism and a category of books unto itself, from Eric Schlosser's *Fast Food Nation* (2001) to Michael Pollan's *Omnivore's Di-lemma* (2006) and beyond.

People began looking at meat in new ways. For decades, America's food system had operated on the principle of opac-ity: steaks and chops magically appeared in supermarkets, and consumers didn't ask many questions. As one industry insider explained, "For modern agriculture, the less the con-sumer knows about what's happening before the meat hits the plate, the better." But scandal threw open a window. When consumers began tracing their pork back to its source, many discovered that they didn't like what modern farming was do-ing to family farms, the environment, public health, and the welfare of pigs.

Before World War II nearly every Corn Belt farmer practiced mixed farming, raising not only hogs and cattle but also corn and soy to feed them. By the late 1960s, however, a livestock expert had already noted "the disappearance of the diversified general Midwestern farm." Farming in the industrial style—with high-priced machinery, hybrid seed corn, and chemical fertilizers, herbicides, and pesticides—expanded crop yields and increased the acreage that one farmer could handle. Many farms disappeared, and those that remained were larger and more specialized. Secretary of Agriculture Ezra Taft Benson infamously offered this advice to farmers: "Get big or get out."

And that's exactly what happened. According to the US Department of Agriculture, there were 3 million farms in the United States raising hogs in 1950. By 2002, that number had dropped to 79,000, a loss of more than 97 percent. In 1950, the average hog farm had 19 animals; a half century later, the average was 766. Much of the consolidation has happened in the last three decades. In 1992 farms with more than 2,000 animals accounted for just 30 percent of the total hog population in the United States. By 2004, 80 percent of hogs lived at these enormous facilities. In 2010, the top four hog producers had captured two-thirds of the market, and the largest—Smithfield Foods—controlled nearly a third of it.

The big producers now control every aspect of that market, from the moment a sow becomes pregnant to the sale of pork chops at the grocery store. Under the traditional system, farmers bred and raised pigs and sold them to pork packers at buying stations or auction barns. Under the new system, many farmers never own the pigs at all but instead act as agricultural foster parents. Much of the work of modern hog raising is done by independent farmers who operate under contract to big corporations, raising pigs according to contractual specifications. The corporation owns the boars that produce the semen, the sows impregnated by that semen, and the piglets born as a result. The company creates proprietary blends of feed and stipulates how the pigs must be fed, watered, and dosed with drugs. The company owns the slaughtering plants, and many times it owns the trucks that deliver packaged meat to retailers. "Vertical integration gives you high-quality, consistent products with consistent genetics," Joseph W. Luter III, chairman of Smithfield, said in 2000. "And the only way to do that is to control the process from the farm to the packing plant."

Consolidation led to geographic shifts. Since the 1820s, most hogs had come from corn-raising states. But starting in the 1980s the industry shifted east. North Carolina, desperate to replace a devastated tobacco sector, welcomed hog farming with open arms. The number of hogs raised there doubled from 1987 to 1992, then doubled again by 1998. It was expensive to ship corn and soy from the Midwest, but overall costs came down thanks to cheap labor, lax environmental regulation, warmer weather (which reduced the need to heat barns), and proximity to eastern markets. Wendell Murphy, founder of a large hog producer, held a seat in the state legislature and unapologetically backed laws friendly to his industry.

Packing plants grew larger as well. By 2006, 95 percent of hogs in the United States were slaughtered at plants that handled more than 1 million hogs a year, and one plant in North Carolina killed 8 million annually. The kill and cut lines moved at lightning speed—1,300 hogs per hour in some cases—and became highly automated. Previously, meatpacking had required expensive skilled labor because there was no way to automate the disassembly of hogs. Tight control of breeding produced not just lean hogs but almost perfectly uniform ones, which allowed machines to do more of the work. The packers hired low-skill, low-wage workers, often immigrants recruited from Mexico and Central America. These laborers had little bargaining power, and their working conditions suffered accordingly: meatpacking soon had one of the highest injury and illness rates of any US industry. It was no accident that the hog industry—like the broiler hen industry—grew most quickly in the South, the region least welcoming to organized labor.

As the hog farms and slaughterhouses grew larger, so did the retailers. In 1997, the four largest grocery retailers held a modest 19 percent of the market. In 2009, the four largest controlled more than half of the market. These large retailers—led by Walmart—demanded meat of uniform quality, in huge quantities, at low prices. To satisfy the grocery chains while maintaining profits, the big pork companies squeezed the least powerful player in the supply chain: the farmers who raised pigs on contract. For every dollar spent on pork, farmers received an ever smaller portion—from fifty cents in the 1980s to less than a quarter by 2009.

Consumers didn't know what percentage of their dollar was making its way to farmers. They only knew they could buy very cheap pork. In constant dollars, the price of pork was 30 percent lower in the 2000s than it had been in the 1950s. In this sense, modern confinement hog farming had been a spectacular success. But cheap meat proved costly, for pigs as well as for people.

After a week of heavy rain in June 1995, the dike failed on an eight-acre manure pond in eastern North Carolina. About 25 million gallons of "red, soupy waste"—a slurry of hog feces and urine—spilled out of the pond, across tobacco and soybean fields, and into the New River, where it killed millions of fish.

"It came through the woods," a neighbor said. "You could see the dark stuff. It made me sick."

The spill exposed a big problem in North Carolina's new hog country. A 250-pound hog excretes 7.8 pounds of feces and 2.65 gallons of urine per day, about four times as much as a human being of equivalent weight. The 60 million pigs in the United States in 1995 produced almost as much waste as the country's 266 million people. Strict rules governed the disposal of human waste; not so the pig waste.

Slatted floors freed farmers from the chore of shoveling manure, but the waste did not disappear. It was flushed into open cesspools thirty feet deep and acres wide. The contents sometimes seeped into groundwater, fouling nearby wells and streams. Assuming the dam didn't break, some of the liquid in the lagoon evaporated, and bacteria partially broke down the waste. What remained was sprayed onto nearby fields. Theoretically it served as fertilizer, but it was applied in concentrations far higher than the ground could absorb and therefore washed into streams. The ponds themselves emitted ammonia, methane, carbon dioxide, hydrogen sulfide, and other noxious chemicals, making life unbearable and sometimes dangerous for those downwind. As one man who lived near a manure lagoon explained, "We are used to farm odors. These are not farm odors."

In 2007 farmers discovered a new problem: for mysterious reasons, manure started foaming and bubbling up through the slatted floors. The bubbles contained gases including methane, a highly flammable by-product of decomposing waste. In 2011, at a farm in northern Iowa, the manure foam exploded, killing 1,500 pigs and seriously injuring one worker.

The problems of waste disposal shaped where hog facilities could be located. In 1997, North Carolina placed a moratorium—it later became a ban—on the construction of new hog lagoons and spray fields. As a result, the industry shifted to the high plains of Texas, Oklahoma, Kansas, and Colorado, where producers found looser environmental regulations and fewer neighbors to complain of the stench.

American taxpayers might not like the sights and smells of confinement operations, but they have been underwriting those operations all the same. For many years, the federal government subsidized the price of corn and soy. A corporation that bought hog feed on the market enjoyed the benefits of that price cut, but a small hog farmer who grew his own corn and soy did not. According to one estimate, federal crop subsidies saved hog producers $8.5 billion between 1997 and 2005. Similarly, the lack of environmental regulations meant that a facility's neighbors and society as a whole absorbed the costs of waste disposal. All told, experts suggested that American taxpayers gave hog producers a subsidy of about $24 on every hog.

And those estimates didn't include other externalities—costs that aren't reflected in the price of a pork chop. Livestock production plays a major role in climate change. Sources of greenhouse gases include forest clearance to grow corn and soy (which releases the carbon stored in trees), feed crop cultivation (which relies on petroleum-intensive fertilizers), and the biological processes of the animals (such as methane in belches and manure). Depending on whose estimate you believe, the global livestock industry generates between 15 and 50 percent of the gases that cause climate change.

Feeding antibiotics to pigs can also lead to drug-resistant bacteria that threaten human health. More than three-quarters of the antibiotics used in the United States go to livestock farms, and according to the FDA, the majority of those drugs are "medically important" within the health-care system. But feeding those antibiotics to pigs plays a role in diminishing their effectiveness in people. When such drugs are used as a feed supplement, antibiotic-resistant bacteria can evolve in animals and then spread to humans, increasing the risk of illnesses that cannot be cured. In response to such dangers, Denmark, a leading pork exporter, banned the use of antibiotics to promote growth, and its pork industry has continued to thrive. Pork producers in the United States have resisted such change, insisting that the agricultural use of antibiotics is safe. The Centers for Disease Control and Prevention (CDC), however, offered an alarming assessment in a 2013 report: "Antibiotic use in food animals can result in resistant Campylobacter bacteria"—a nasty bug that can cause bloody diarrhea and temporary paralysis—"that can spread to humans." The report also warned that drug-resistant salmonella followed a similar path from animals to humans. The CDC concluded, "The use of antibiotics for promoting growth is not necessary, and the practice should be phased out." The FDA in 2013 issued a new policy with precisely that aim, but critics suggested that loopholes would allow current practices to continue.

Pigs have shouldered the most direct costs of modern hog farming. The pork industry, not surprisingly, claims that confinement barns are perfect for the animals. "They love it," the president of one pork corporation has said. "The conditions that we keep these animals in are much more humane than when they were out in the field." The evidence suggests otherwise.

Hogs in confinement endure a litany of horrors. They spend their lives standing over septic pits, and ventilation systems are often used sparingly to keep heating and electricity costs down.

Workers who enter the barns sometimes don respirators, a convenience not available to the thousands of pigs who breathe the air constantly. "Acute and chronic infections of the respiratory tract in pigs are common," according to *Biology of the Domestic Pig*, a standard reference work. "Most are related to . . . inadequate ventilation, improper temperature and humidity, poor nutrition, and high levels of irritant gases such as ammonia."

Hogs also develop behavioral problems that the industry refers to as "vices." One vice is a tendency to bite the tails of other pigs, which can lead to infection. Market pigs are juveniles, just five or six months old when slaughtered, and are as intelligent, playful, and curious as puppies. On pasture or in barns, they spend much of their time exploring with their mouths and snouts—rooting, chewing, poking, exploring. But in a confinement facility with metal bars and concrete floors, such instincts find an outlet on the only soft surfaces around—other pigs. "Without malleable substrates to chew, pigs direct what appears to be inquisitive or exploratory behaviors toward other pigs and are rewarded by movement of the tail and perhaps the taste of blood," according to a team of university livestock experts. The industry's solution to this problem is to cut off the tail, leaving only a one- or two-inch stub so sensitive that a pig would attack any other pig that dared nip it.

Sows suffer particularly shocking conditions. For nearly all of her life, each sow is allotted a space only slightly bigger than she is. These cages, the industry says, prevent fighting among sows and keep them from crushing their newborn piglets. These claims are true. But it is also a matter of economics. "Once a building is built, the most efficient use of space will be to have as many units of production in that building as is possible," one industry expert explains. "The sow crate allows the greatest number of breeding sows in a given building space."

Sows in modern hog facilities spend nearly all of their lives in two-by-seven-foot crates. The system keeps them from fighting each other and maximizes space in barns, but it also prevents them from walking or even turning around. The sows' suffering has prompted campaigns to ban the use of crates. (Courtesy Humane Society of the United States)

But efficiency causes problems for these "units of production." From standing so long on hard floors, sows develop foot and hoof lesions, a problem that afflicts as much as 80 percent of the sow herd. Behavioral problems are just as widespread. Sows cannot walk or even turn around. They cannot groom themselves or interact with other sows. They can only stand up, lie down, eat, and defecate. With no outlets for natural instincts, they develop what is known as "stereotypic behavior," such as chewing on the bars of the stall until their mouths bleed. Sows preparing to give birth, who by instinct build nests for their piglets, cannot do so because they have no access to straw. But they keep trying: they swing their heads up and down, back and forth, as if building nests with invisible straw.

Animals suffered, of course, even under more traditional forms of agriculture. Sows accidentally crushed their piglets and sometimes ate them. Pigs on pasture froze to death during cold snaps. Parasites and infectious diseases were rampant, and hundreds of thousands of hogs died terrible deaths from hog cholera, dysentery, and influenza. Such suffering, while not uncommon, indicated that something was wrong. The problem with confining pigs is that cruelty is built into the system. Tail docking, gestation crates, slatted floors bare of straw: all have been shown to cause suffering. Yet all are standard operating procedure.

The pig's plight is well understood but tends to be ignored or rationalized away. "Contemporary swine production systems may create frustration in pigs because of the lack of materials or opportunities to perform certain behaviors," a group of livestock specialists admit. But then they add a hopeful note: "Pigs display a remarkable ability to adapt their behavior to the husbandry setting." That is true. If the history of the pig tells us nothing else, it is that these animals can make do just about anywhere, eating just about anything.

Thus far, however, pigs have failed to adapt to tiny crates, crowded pens, slatted floors, and ammonia-saturated air. Some needs—to build a nest, to root, to nuzzle—should not be denied.

EIGHTEEN

"BACK TO THE START"

B ernard Rollin, an ethicist at Colorado State University, once
gave a speech to a group of pork producers in Ontario,
Canada. He was nervous. He had criticized industry practices
in a column he wrote for the *Canadian Veterinary Journal*, but
he had never spoken directly to a group of hog farmers. Cattle
ranchers, he had found, "cared deeply about how they managed an animal, even if it meant losing money or sleep treating
a sick creature." But cattle raising had preserved many of its
traditional methods, while hog production had been thoroughly
industrialized.

Rollin spoke outdoors to two hundred hog farmers gathered
around picnic tables. He pointed out that society was becoming
increasingly concerned about the treatment of animals in zoos,
in circuses, and on farms. He called confinement operations
"exploitative"—because they treat animals as unfeeling meat-
producing machines—and made the case for an old-fashioned

ethics of animal husbandry, which insists that the human partnership with livestock carries moral responsibilities. Animals suffer and deserve compassion, even if that compassion hurts the bottom line. In his talk to the farmers, Rollin says, "I beseeched them to look into themselves, examine what they were doing and see if it accorded with their own ethics."

He stopped talking. There was silence for a moment, then applause. Finally one farmer in the audience jumped atop a picnic table. "I have been feeling lousy for fifteen years about how I raise these animals," the man said. "I am pledging to tear down my confinement barn and build a barn I don't have to be ashamed of."

If that story sounds a bit like a television commercial, that's because a restaurant chain later created a commercial on a similar theme. Chipotle, which claims to use pork only from pigs raised on pasture or in pens with deep beds of straw, released a two-minute animated film titled *Back to the Start* in 2011. The story, which has no dialogue, begins with a young farmer, joined by his wife and baby, working a bucolic small farm with pigs on pasture. He then builds an enormous factory where his pigs are pumped full of feed and drugs before being packed into boxes and driven away on big trucks. Finally, as an old man with a grown son, the farmer dismantles the factory and turns it back into a farm, with pigs on pasture once again. On the soundtrack, Willie Nelson sings a Coldplay song: "Science and progress don't speak as loud as my heart / I'm going back to the start."

It's a heartwarming tale and not untrue: in America and Europe, a growing number of farmers are embracing more traditional forms of agriculture. But that's only a small slice of the story. In the developing world, especially China and Brazil, farmers are moving in the opposite direction: they are giving up old-fashioned ways and raising pigs in confinement barns because they can make a lot of money doing so. In only the latest controversy in the long history of swine, some people are tearing down confinement barns out of ethical concerns as others raise them up for financial reasons. It's a question of values and prices: How much is a happy pig worth?

Westerners, on the whole, eat slightly less meat than they used too, but elsewhere the appetite for meat is growing quickly. Meat consumption in the developing world more than doubled between the 1960s and 2000. China and Brazil accounted for nearly all of that growth, and demand is likely to keep rising in the next few decades.

Production has risen to meet this global demand. In developed countries—the United States, the nations of western Europe, and Japan—the amount of meat produced rose only slightly between 1980 and 2004. During that same period, meat production in the developing world increased by 300 percent, with pigs and poultry recording the biggest gains. The same calculations made in Iowa in 1960 are being made in Brazil today, as grassland and rain forest are destroyed to make way for feed crops: cows need to spend a large part of their lives grazing on grass, but that cow pasture can be plowed under to plant corn and soy as feed for pigs and chickens, a process that produces more meat on less land.

Thanks to genetically modified seeds, fertilizers, pesticides, and herbicides, crop production has boomed, and feed prices have dropped: in constant dollars, corn and soy cost about half as much as they did in 1961. Cheap grain flows around the world on ships, destined for the stomachs of pigs and poultry. Chicken production has risen fastest because religious prohibitions do not limit its consumption, but pork production has nearly kept pace, thanks largely to China.

China has a pork-based cuisine, a population of 1.3 billion people, and a growing middle class that now earns enough to eat meat regularly. That's why the government maintains a strategic pork reserve, consisting of both live pigs and frozen pork. It functions much like the federal oil reserve in the United States: in times of shortage, China releases pork onto the market to stabilize prices and prevent social upheaval.

Though China is still largely self-sufficient in pork, this won't stay true for long as demand increases. Part of the solution lies abroad: in 2013 the Chinese meat producer Shuanghui International paid nearly $5 billion to buy Smithfield Foods, America's dominant pork producer. At the same time, Archer Daniels Midland and other commodities firms have spent more than $10 billion buying up grain traders in the United States, Australia, and elsewhere, so that they can sell crops to China. Once a leading exporter of soybeans, China is now the leading importer, buying more than half of the world's supply. Those foreign soybeans do not get turned into tofu; they become pork.

Chinese pork production is in the midst of vast modernization. As recently as the 1980s, small farmers who raised fewer than five pigs a year produced more than 95 percent of the China's pork. Two decades later, that figure had dropped to less than 25 percent. The small farmers are efficient, relying on thousands of years of agricultural wisdom, but they can't keep up with the nation's meat hunger. Pigs, once fed on weeds and rice bran, now eat imported corn and soy, mixed with supplements and antibiotics. Their manure, once used to fertilize rice paddies, now poses a pollution hazard. Genetic diversity—represented by more than one hundred local pig breeds—is being displaced by

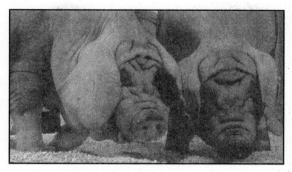

As China grows richer and hungrier for meat, it is abandoning traditional small-scale production and embracing confinement barns. Traditional Chinese pig breeds, like these Meishan, are giving way to hybrids indistinguishable from those found in the United States. (Keith Weller; courtesy of the US Department of Agriculture)

Western hybrids with the same genetic lines that can be found in North Carolina and Brazil.

In three decades, China has experienced a transformation of diet and agriculture that took a couple of centuries in the United States and Europe. Western nations, meanwhile, are entering a new phase of their agricultural history, one in which consumers seek a balance between an appetite for cheap meat and the qualms of conscience—a balance that some try to maintain by understanding the innate needs of pigs.

In the late 1970s Alex Stolba and David Wood-Gush, scientists at the University of Edinburgh in Scotland, built a multi-acre "pig park" that included a small pine forest, gorse bushes, pastures, a stream, and some boggy ground. Then they bought standard commercial pigs—one adult male, four adult females, a juvenile male, and a juvenile female—and set them loose. The scientists studied the pigs as if they were little-understood, exotic creatures, which, in a sense, they were. Although pigs had lived in great intimacy with humans for 10,000 years, much basic information about their instincts had gone unrecorded.

The researchers put a scientific polish on facts swineherds had known for millennia. The pigs at the Edinburgh pig park were most active in morning and evening, resting at night and in the middle of the day. Most activity—rooting, eating, nuzzling—involved the snout, the pig's key interface with the world. On some days pigs spent more than six to eight hours rooting and foraging. The pigs rubbed their faces against tree trunks, marking them with scents from glands in the face, and other pigs later visited the trees to sniff out the messages left behind, much as dogs do with markings left by urine. The pigs also formed complex social bonds: females from the same litter tended to stick together long after weaning, and piglets maintained bonds with their mother even after she'd given birth to her next litter. Unknown pigs introduced into the park found themselves harassed and excluded for months before finally being allowed into the herd.

The pigs slept in a communal nest, which each pig would help to refresh nightly with new mouthfuls of grass and straw. They built several nests over the course of the study, and all had certain things in common: the nests were distant from the feeding sites, offered a wide, clear view of any approaches—the better to spy oncoming predators—and were protected on at least one side from the chill wind. Upon first arising in the morning, the pigs walked at least twenty feet away before urinating or defecating.

In the hours before giving birth, a pregnant sow would choose her own private nest site, far from the communal nest, and then gather material, picking up sticks and straw, cropping grass with her teeth, and arranging it for warmth and comfort. After farrowing, she would not allow any other pigs to approach for several days. With gentle grunts and nudges from her snout, she assembled the piglets on one side of the nest, then moved a few feet away before lying down. Only then would the piglets move alongside her to nurse. Piglets ventured out of the nest after five days or so, and they weaned themselves at three or four months.

After their pig-park study, Stolba and Wood-Gush concluded that the pigs' behavior "resembled that of the European wild boar." Domestic pigs, that is, reverted to their wild ways once given a bit of room to roam. Given what we know about feral pigs, this is not surprising. A few million years of evolutionary history easily override a few thousand years of domestication.

The pig park and many behavioral studies that followed showed just what was wrong with confinement barns. Pigs need materials to manipulate—wood chips or straw at the very least—or they will use their snouts and teeth to hurt themselves or each other. Keeping pigs in family groups is wise because they're more likely to fight with outsiders. The sow's nest-building instinct is strong and should not be denied. Studies also determined that there is a genetic component to maternal behavior. This should shock no one: mothers who don't crush their piglets will pass along their genetic materials more successfully. If breeders selected for maternal abilities as well as rapid weight gain, farrowing crates would not be necessary.

It was no coincidence that the pig park study took place in the United Kingdom, which since the nineteenth century has been at the forefront of the fight for animal welfare. In the modern era the seminal work has been Ruth Harrison's 1964 *Animal Machines*, a graphic critique of industrial livestock farming that galvanized Great Britain's animal rights movement, much as Rachel Carson's *Silent Spring* helped create modern environmentalism in the United States. In response to *Animal Machines*, the British government formed the Brambell Commission "to enquire into the welfare of animals kept under intensive livestock husbandry systems." The inquiry uncovered appalling conditions and declared that, at the very least, farm animals should have enough room to "stand up, lie down, turn around, groom themselves and stretch their limbs."

These modest requirements became known as the "five freedoms," a sort of bill of rights for livestock. A group called the Farm Animal Welfare Council later revised the five freedoms, arguing that animals should enjoy "1. Freedom from hunger and thirst. 2. Freedom from discomfort. 3. Freedom from pain, injury, or disease. 4. Freedom to express normal behavior. 5. Freedom from fear and distress." These recommendations carried no legal weight, but they proved influential throughout Europe.

The European Union (EU) has been the most powerful agent of change in the conditions under which modern livestock live or die. In 1997 an EU veterinary committee issued a 190-page report titled "The Welfare of Intensively Kept Pigs." It was a measured document, sensitive not only to the humane treatment of pigs but also to the economic interests of farmers. It nonetheless came to damning conclusions about pig farming, and as a result the EU issued mandatory new standards. Europe required that pigs have access to larger areas of solid flooring for more comfortable lying and standing and that the gaps in slatted floors be narrowed to minimize foot injuries. Sows had to have constant access to materials for rooting. The farthest-reaching provision banned the use of gestation crates: instead of tiny individual stalls, all sows had to be housed in groups. Canada, too, has since ordered that gestation crates be phased out.

Promoting farm animal welfare has proved more difficult in the United States, a country generally suspicious of government regulation. Private groups, however, have had some success, especially with efforts to restrict gestation crates. In 2008 the Pew Charitable Trusts, a prominent nonprofit, issued a harsh report on modern livestock farming that called for banning gestation crates because they "restrict the natural movement and normal behaviors of animals." The Humane Society became the most active campaigner, releasing detailed reports and alarming photos and videos. "Gestation crates are a real problem," said animal scientist Temple Grandin, who is widely respected within the livestock industry. "Basically you're asking a sow to live in an airline seat." The images of large animals jammed into tiny pens called to mind earlier activism against crated veal calves—a campaign so successful that veal became almost taboo. A number of US states have banned gestation crates, but none of them are major hog producers.

Change in the United States is less likely to happen through government intervention than through corporate buying power, as businesses respond to growing consumer repulsion over gestation crates. Burger King in 2007 became the first major restaurant chain to announce it would require its suppliers to phase out the crates, and many others followed suit, including McDonald's, Denny's, and Cracker Barrel. Retailers and wholesalers—Kroger, Costco, Sodexo, and more—announced similar moves. Under pressure from these major buyers, Smithfield and its rivals Hormel and Cargill announced that they would shift away from gestation crates, but only very gradually. Early in 2014, Smithfield promised to be crate-free by 2022, to give its farmers time to make the switch.

Industry spokesmen tend to respond to animal welfare concerns with a mix of resignation and contempt. They insist that

gestation crates are humane but agree to comply with their customers' demands: if McDonald's wants to buy crate-free pork,
then Smithfield will produce it. "Their feelings aren't rational,"
an executive with an American meat company said of consumers, "but they are important." In the words of one European
official, "One must concede the right of choice to the buyer,
even when such a choice involves an element of gimmick and
charade."

Thanks to legal changes in Europe and consumer pressure
in the United States, confinement hog operations are making
minor modifications to their current methods. Instead of being
contained in gestation crates, more sows are living in group
pens. Instead of bare, slatted concrete floors, they have areas
of solid flooring and some straw. The changes are significant,
but for some they aren't enough. The EU, for instance, has a
stricter standard for "organic" pork, which prohibits tail docking, teeth clipping, and growth-boosting drugs and also requires
organically grown feed and access to outdoor space. Producers
in Denmark have created a special category called the "welfare
pig"—a phrase that would carry rather different connotations in
the United States—in which the animals are kept in conditions
that split the difference between organic and conventional. Similar standards in the United Kingdom qualify a pig as "Freedom
Food," a labeling effort that has found broad success, including
a commitment from the British arm of McDonald's to buy only
pork that meets the standard.

The American Humane Association adopted the British
Freedom Food standards and created the American Humane
Certified label, part of a growing trend in the United States.
Other private programs include Animal Welfare Approved and
the Global Animal Partnership, which is closely linked to the
upscale grocer Whole Foods. The meat case at every Whole

Foods store grades each piece of meat on a five-step scale, pro-gressing from "no cages, no crates"—the grocer does not sell any conventionally raised pork—to "enriched environment," to "enhanced outdoor access," to "pasture-centered," to "ani-mal-centered." The US Department of Agriculture maintains its own standards for "organic" pork, which prohibits antibiotics and requires that animals eat organically produced feed and have access to the outdoors, and "natural" pork, which refers only to a lack of artificial additives and has nothing at all to do with the way the animal was raised.

These certifications serve as marketing tools: they allow consumers to buy pork based not only on price and quality but also on the animal's living conditions. Increasingly, consumers are demanding humanely raised meat. And in the United States, no one has more successfully filled this niche than Niman Ranch.

Paul Willis, who created Niman Ranch's pork operation, grew up on a farm in central Iowa, went off to college, worked for the Peace Corps in Africa, and then returned to the family farm. He liked pigs and wanted to watch them playing in fields, so he didn't build the new-style barns. Confinement farming is "not better for the animals, or for the farmers, or for consumers," he said. He kept to the traditional Corn Belt way, growing his own corn and soy and keeping his pigs on pasture. Willis's pigs had virtually nothing in common with conventional pigs until the end of their lives: they were sold to a Hormel plant, where they commanded the same price as hogs from confinement barns.

In the mid-1990s Willis met Bill Niman, who had been sell-ing grass-fed lamb and beef raised by California ranchers, and the two struck up a partnership. Willis recruited like-minded farmers in Iowa, who at the time were more than willing to

Concern for the humane treatment of livestock has prompted the creation of organizations such as Animal Welfare Approved, which certify that private farms are raising animals according to certain standards. Promotions carrying pastoral images of pigs help persuade consumers that such pork is worth its higher price tag. (Courtesy Animal Welfare Approved)

listen. In 1998 and 1999 prices for conventional US hogs fell to historic lows, and farmers turned to Niman Ranch as an alternative buyer. Willis paid a premium for pork raised according to his standards and sold it to restaurants and grocery stores, as well as directly to consumers via mail order. Eventually, his network grew to hundreds of farms. "All our hogs are raised outdoors or in deeply bedded pens, where they are able to express their natural instinctive behaviors, like rooting and roaming," Niman Ranch explains on its website. The pigs "express exceptional mothering abilities"—hence don't need farrowing crates—eat a "100% vegetarian diet," and are "never given hormones or antibiotics."

Niman Ranch allows its pigs to be raised indoors so long as they have plenty of straw. Other, smaller producers insist that all pigs have access to pasture. EcoFriendly Foods, for example,

advertises that all of its pigs can give free rein to their omnivorous appetite, eating "insects, larva, crayfish, lizards, medicinal plants and roots, grasses, seeds, nuts, legumes, fruits,
etc." Those foraged foods are supplemented with "a mixture of
locally-sourced grains, legumes, root crops, nuts, fruits, berries,
kitchen scraps, and salt and minerals."

Consumers and farmers are realizing that raising old-
fashioned hogs in old-fashioned ways pays dividends in taste
as well as in animal welfare. Niman Ranch hogs are a cross of
the Duroc, Chester White, and Berkshire breeds, while other
farmers raise purebred Berkshire, Hampshire, Gloucester Old
Spot, Tamworth, or American Guinea swine—traditional types
now referred to as "heritage breeds." Such pigs are adapted to
life outdoors, and their meat, rather than being stripped of its
fat, is well marbled and delicious. Because a pig's flesh expresses
its diet, an animal that lives on pasture and eats apples, acorns,
and grubs will produce a more complex pork loin than one fed
just corn and soy. In addition, muscles that get exercise develop
better flavors than those that don't. A chicken's legs, which work
to support its weight, have more flavor than its breast muscles,
which are never used to fly. For the same reason, pigs that run
around on pasture taste better than those packed into tiny pens
their entire lives.

Such pork is becoming more popular, but the competition
does not make Smithfield quake with fear. "Niche meat" is
the industry term for animals raised in such alternative systems,
and the niche is small. A study at Iowa State University estimated that 500,000 to 750,000 niche-market pigs were slaughtered in 2005, compared to more than 100 million conventional
pigs. More recent statistics are not available, but even if niche

pork production has quadrupled in the past decade, it would still account for just 3 percent of the American pork market.

The key sticking point, as ever, is price. A consultant for Tesco, the huge British supermarket chain, experimented with a modified confinement system and found that it increased the cost of meat by 30 percent. "Good welfare means that the base price of pork will inevitably rise," she explained. In fully pastured systems, with slower-growing heritage breeds, the prices climb ever higher.

Chuck Talbott, who has a PhD in animal science and now raises pigs near Charlestown, West Virginia, has seen firsthand the trade-offs between price, ethics, and taste. Talbott and two partners run a high-end curing operation called Woodlands Pork. The pigs—a blend of heritage breeds with some wild boar mixed in—spend their final months in the woods eating acorns and other wild foods. Then his partners process and cure the meat, producing a pricey American ham to rival the best from Europe.

Talbott explains the dilemma of modern pork in these terms: "You've got the rich farmers feeding the poor families, and the poor farmers feeding the rich families." Which is to say, big corporations run by wealthy executives sell bacon for $3 a pound at Walmart, while struggling small farmers offer it at farmers' markets for four times that price. We see, once again, the pig's ability to divide.

VIRTUOUS CARNIVORES

The way pigs are raised has changed frequently over the centuries, but attitudes toward pork have remained largely the same. A 2013 food industry study found that 92 percent of Americans ate beef at least once a week, while only 64 percent ate pork. When presented with a salad or burrito that came with a choice of protein, a third of restaurant customers chose beef, another third opted for chicken, and 12 percent picked fish. Then came pork, at 9 percent, beating out only the vegetarian options. Recent figures from the US Department of Agriculture show that rural dwellers eat more pork than urbanites, southerners eat more than northerners, and the poor eat more than the rich. The situation, in other words, has remained unchanged since the 1950s, despite the best efforts of the Porkettes and the industry's "The Other White Meat" campaign. And these eating patterns are not unlike those of 3,000 years ago, when the ruling classes of Mesopotamia and Egypt stopped eating pork.

In the United States today, as in the ancient Near East, many people, for many reasons, are avoiding pork. Some of those reasons—status, religion, disgust with the pig—are ancient. But there is also a new reason, and it relates to the brain inside the pig's head. In 2013 the activist group Mercy for Animals released footage recorded inside a Minnesota hog barn that showed runt piglets being grabbed by their back feet, slammed onto the concrete floor, and tossed into an overflowing bin of dead and dying piglets. This sort of footage has become disturbingly familiar, but one thing set this video apart: James Cromwell, human star of the movie *Babe*, appears onscreen as the narrator. Cromwell's appearance draws an unspoken contrast between brutalized pigs destined for slaughter and the adorable pig named Babe who avoids that fate by learning to herd sheep. In the movie, Babe controls the flock not as the dogs do—through threat of violence—but by learning the sheep's language and making polite requests. Babe treats sheep not as dumb brutes but as smart, sensitive creatures. Babe's owner, played by Cromwell, follows a similar learning curve: he slowly comes to recognize intelligence, even in a pig.

The world learns this lesson again and again but never fully absorbs it. Though the pig's cleverness has been noted at least since Roman times, it seems that each era must make the discovery anew. Recently, however, as cognitive and behavioral scientists have confirmed the pig's talents, the contrast between the animal's intelligence and its living conditions has become harder to ignore. Critics don't have good things to say about any type of industrial livestock farming, but they tend to agree that hog production is the worst. Pigs are the most intelligent, the most human-like, of the farm animals, but they are kept tightly caged and never see daylight from birth to slaughter. British chef Hugh Fearnley-Whittingstall calls pigs "the most abused of all our farm animals" and says that when he looks at supermarket meat cases, he sees "strip-lit morgues for millions of miserable pigs." Activist Gail Eisnitz, who has investigated all types of animal cruelty, found herself most troubled by the hog barns she'd visited, "especially seeing all those caged sows—waving their heads, chewing the air," she said. "It's like an insane asylum." These sows, Jonathan Safran Foer writes in *Eating Animals*, "often will develop quarter-sized, blackened, pus-filled sores from chafing in the crate" and "must lie or step in their excrement to force it through the slatted floor." Pig feed, Matthew Scully points out, often contains the processed remains of other pigs. That means that sows, after giving birth to eight or more litters over a period of four years, will be "slain and rendered into feed for their progeny to eat."

These writers hit many of the same notes used to criticize pigs for thousands of years: pigs wallow in filth and eat disgusting things. But the target of outrage has shifted. Unlike Saint Francis and other medieval moralists, modern critics don't blame the pigs. They blame, instead, the corporations who raise them this way. "The pigs are treated like shit," chef David Chang said. "I always say that big pork companies are no different than big tobacco companies." Tom Colicchio, another prominent chef, echoed that view, describing large pork producers as "disgusting."

These critics are making a plea for compassion. But they are also creating a new language of pig disgust—one that has nothing to do with pigs being gluttonous or lustful, garbage eaters or trichinosis carriers, and everything to do with the way that people treat pigs.

Many people see problems with the way hogs are raised, but they tend to disagree on solutions. A few animal scientists have proposed—in a *New York Times* op-ed, among other places—breeding a new type of pig with compromised mental abilities, a genetically lobotomized animal that could endure industrial conditions without stress. This idea, thankfully, has made little headway. Most American hog producers, under pressure from consumers, have started to follow the example of those in the European Union: they are making incremental adjustments—group housing rather than gestation stalls, straw bedding rather than bare floors—to address the worst problems of confinement. Another option is to refuse to eat meat altogether. That's the position taken by Cromwell, who in the Mercy for Animals video advocates "leaving pork off your plate and adopting a vegetarian diet."

Other producers choose a middle path. They are disgusted by industrial pork but entranced by pigs raised on pasture, or in the woods, or at the very least in a barn on a deep pile of straw. They are the new pig haters and the new pig lovers. They reject the swine of modern agribusiness but embrace the breeds they find by digging deep into the history of swine. In addition to working with the more familiar heritage breeds like the Berkshire and Tamworth, a few are experimenting with exotic varieties such as Mangalitsa, a Hungarian lard-style pig with curly hair, and Ossabaw, descended from a population of feral pigs that survived on a Georgia coastal island for hundreds of years. And thousands of small growers are adding pigs of various breeds to their mixed farming systems because it makes sense, just as it has for thousands of years: they buy a couple of feeder pigs for $50 a head, raise them on whatever waste they have on hand—damaged vegetables, whey, stubble in the fields—finish them on barley or corn to harden the fat, and then sell the pork at farmers' markets.

Chefs have been the key allies of these old-fashioned farmers. The Roman historian Livy noted in the second century BC,

"The cook, who had formerly had the status of the lowest kind of slave, first acquired prestige, and what had once been a service-industry came to be thought of as an art." That's not a bad description of what has happened over the past two decades, as cooking has moved into the mainstream of popular culture. And pigs have been one of the prime beneficiaries.

Modern chefs, like those in ancient Rome, love pork—an inordinate number have pig tattoos on their shoulders—and they are enamored of exotic preparations. No restaurants have resurrected the Roman recipe for roast udder of lactating sow, but chefs have cooked guinea hen in pork bladder, while roasted pig snout and fried pig ears have become almost common at farm-to-table restaurants. At the New York restaurant Maialino, the menu featured half a young pig's head atop a salad. It sold well—"perhaps because of the shock value," chef Nick Anderer speculated.

Pork has always had higher shock value than other common meats. That's partly because the pig's feet, ears, and other odd parts find their way onto the plate fairly often. Pork also shocks simply because it is fatty, offering forbidden pleasures that flout mainstream health guidelines. The pork industry has played both sides of this fence. Even as pork marketers flogged lean pork chops in the 1990s, they also urged fast-food restaurants to top their burgers with precooked bacon slices. This campaign, executed with far less fanfare than "the other white meat," ultimately enjoyed more success: sales of lean pork stayed flat, but demand for pork bellies surged. Burgers began to seem naked without a slice of bacon or two—or even six, as on the Wendy's Baconator. By the 2000s bacon had developed its own cult whose adherents bought bacon-themed toilet paper, bandages, and water bottles (with the slogan "Bacon Squeezins. Refreshing!"). A couple of cooks in Kansas City found Internet fame by cobbling together a monstrosity called the "Bacon Explosion": slices of bacon woven into a mat, covered with sausage, and rolled into a log. A photo of a young women wearing a homemade "bacon bra"—strategically draped raw slices—showed that the link between pork and sex, which dates back to ancient Greece, remains unbroken. Dozens of cities hosted bacon festivals. And in perhaps the surest sign that the trend had gone mainstream, Denny's advertised a "Baconalia," although Bacchus—Roman god of wine celebrated at the riotous Bacchanalia—would have been disappointed to learn that Denny's most decadent offering was a maple-bacon sundae.

In the bacon festival—and its more venerable cousin, the barbecue festival—one can hear faint echoes of Renaissance London's Bartholomew Fair. Then, the delight was rooted in roast pork's symbolic association with gluttony and lust. Now, the transgressive pleasure derives from thumbing one's nose at fat-fearing dietary scolds. Josh and Jessica Applestone—authors of *The Butcher's Guide to Well-Raised Meat* and founders of Fleisher's, a Hudson Valley butcher shop that has become something of a finishing school for new-school butchers—call pork "the sexiest of all meats." It's an indulgence, they write, that prompts their customers to sidle up to the counter with a "furtive look" and whisper, "Have you got any lard?"

But one must indulge carefully. The title of the Applestones' book contains a warning: most pork—including that featured at bacon and barbecue festivals—is not "well-raised." Ethical dining requires that the hedonist and the puritan embrace. This isn't the first time people have seen the need to change their dining habits for the good of society: in ancient Rome and medieval times, authorities passed sumptuary laws to try to limit excessive feasting, which was seen as detrimental to good social order. The new pig lovers have seized the moral high ground for themselves. For many consumers, it's acceptable to spend a great deal on pork and to indulge in exotic preparations as long as the pig is treated well before dying at a welfare-certified slaughterhouse.

The world of humane farming has produced a new tribe: the virtuous carnivores. "We have been raising happy, healthy pigs since 1994," a farm in Pennsylvania claims on its website, targeting exactly this demographic. Such pigs are easy on the palate as well as the conscience. EcoFriendly Farms reduces it to an equation: "a happy pig = a tasty pig."

Every tribe defines itself by comparison to others, and the virtuous carnivores are a tiny group compared to the vast army who choose meat based on price. But this new group wants to recruit, not exclude. This is partly out of concern for the animals and partly out of self-interest: as more people eat well-raised pork, economies of scale will cause prices to fall. If public concern drives further agricultural reforms—better animal welfare, stricter environmental controls, less use of antibiotics—factory farming will become more expensive, and the price gap between Sam's Club pork and farmers' market pork will narrow, a development that will benefit pigs and people alike.

In the spring of 2012 I got a glimpse of this new meat economy at a workshop called "Advanced Meat Curing" taught by chef Craig Deihl at a community college in Asheville. Sixteen of us—mostly chefs and butchers, as well as one farmer—sat on high stools gathered around a stainless steel table bearing the carcass of a two-hundred-pound pig. Deihl sawed through the pig's neck and then boned out the head with speed and precision. When he was finished, the table held a mostly bare white skull and a single piece of skin and flesh—complete with ears and snout—that might have been mistaken for a gruesome Halloween mask. The entire piece would be rolled, tied, cured, and slow-cooked to create *porchetta di testa*, several pounds of amazing meat derived from a part of the beast that's often rendered for dog food.

Deihl, executive chef of Cypress restaurant in Charleston, South Carolina, has become a cured meat guru. He built a special temperature- and humidity-controlled room for aging and started an "artisan meat share," serving up monthly portions of local cured and fresh meats to subscribers. This is not Walmart fare. Buying into the share program cost $50 a month, and Deihl and the others chefs at the workshop work at restaurants where appetizers ran $15 and entrees twice that. They talked of cold-smoke guns, fermentation cultures, immersion circulators, and Cryovac machines. Their world is high-tech and high-end, dedicated to pleasing the palates of the upper crust.

But there was another goal as well. The meat-curing workshop was part of a larger conference on "whole-animal utilization," hosted by a state-sponsored group dedicated to helping small-scale farmers. Casey McKissick, who organized the conference and also raises hogs and cattle in Old Fort, North Carolina, told me what inspired the event: "Chefs have to understand that if they want to take on local meat in a big way, they have to take on the whole animal." That's because the cuts most grocery stores and restaurants sell—steaks, roasts, rack of lamb, pork chops—make up only a small portion of the carcass. Small meat producers don't slaughter enough animals to satisfy a restaurant that wants to serve eighty pounds of center-cut pork chops a week. But if more chefs bought a whole hog, cut it themselves, and learned to cook, cure, and sell every part of the animal, there would be more well-raised meat to go around.

The market for this type of meat is growing, but that doesn't mean it's an easy way to make a living. Heritage-breed pigs grow more slowly and require more feed than the genetically engineered factory-farm variety. It's more expensive to raise pigs by the dozens rather than the tens of thousands, to keep them on pasture rather than in confinement barns, and to slaughter them at small plants rather than enormous factories. Until virtuous carnivores get their way and well-raised meat becomes mainstream, the only way to keep prices reasonable is to wring a profit out of every last part of the beast.

Once upon a time, snout-to-tail eating was a matter of survival: meat was too precious to waste even the tiniest scrap. For small farmers today, using the whole animal has some of the same urgency. Only by allying themselves with chefs and adventurous eaters can they make enough money to stay in business.

That was the point of the curing workshop. Deihl was teaching people how to make a profit from animal parts that often go to waste. If a chef boils up the head for headcheese, cures fatback into lardo, and gathers up every scrap of meat and fat to be ground into bratwurst or andouille, the restaurant makes more money and can afford to pay farmers to raise pigs in ways that are good for the land, the animal, and the soul.

This new pork economy depends upon willing customers. "You may end up paying twice as much, or even more, for pork," Fearnley-Whittingstall writes in his *River Cottage Meat Book*. "But given that the end product will be infinitely better eating, I would call that a bargain." A bargain, though, is in the eye, and the wallet, of the beholder. The farmers' markets and upscale grocers that typically sell well-raised pork attract people from all walks of life, but those who shop there tend to be wealthier than the general population. A 2007 survey conducted at Oklahoma State University found that only about 30 percent

Snout-to-tail dining, once a necessity for hungry peasants, has reemerged in the realm of high-end dining, as restaurants plate up tongues and tails and turn every stray scrap into sausages—such as these made by Craig Deihl of Charleston, South Carolina. This style of cooking and curing helps farmers cover the higher cost of raising pigs on a small scale. (Courtesy Cypress Restaurant)

of American consumers were willing to pay more for humanely raised meat. The majority of people, the study concluded, "truly value low meat prices more than animal happiness."

Nonetheless 30 percent represents a large market, and it's growing. If it's to overtake the portion of the market that still prioritizes price, however, some things will have to change.

Modern consumers have become accustomed to factory-farmed pork, and for good reason. In historical terms, Americans now spend a tiny portion of their income on food—about 14 percent, compared to 32 percent in 1950 and 43 percent in 1900. We've gotten used to low food prices, which have also freed up funds for the consumer spending that drives Western economies. And we've even gotten used to the taste: cookbook authors might scoff at the flavor of supermarket pork, but most of us don't object. The bacon is still salty and fatty and crisps up beautifully. The pork shoulder and ribs, after spending eight hours in a smoker at the local barbecue stand, are still tender and delicious. The loin, chunked and seared and doused in sweet-and-sour sauce at a stir-fry restaurant, provides the satisfying chew and stomach-filling properties that we desire in meat.

For those of us who don't make our living hosting cooking shows or writing cookbooks, commercial pork tastes good enough. Before upgrading our meat, we will have to develop more discerning palates, or decide to seek prestige in pricier pork, or grow disgusted by factory-farmed meat, or become convinced that the environmental, safety, and animal welfare benefits of humanely raised pork are worth the extra money.

If consumers make these changes, the pork industry will change with them. It's certainly shown the ability to change before. According to John McGlone, an agriculture professor at Texas A&M University, a hog farmer surveying the last fifty years might have this to say about American consumers: "They wanted inexpensive meat. We gave it to them. Then they wanted meat with less fat. We gave it to them. Then they wanted it to taste good, not pollute the environment, be safe to eat, and be good for the animal's welfare." That statement is presented in the voice of a farmer exasperated with the shifting views of a fickle public. But is it really too much to ask?

Twenty-first-century consumers, who have bent so many other industries to their collective will, should be able to change this one too. On a trip to a grocery store or a restaurant, we should be able to buy meat that tastes good, doesn't pollute, doesn't make us sick, and comes from an animal that was treated well—from a pig that lived like a pig. And to support those goals, all we have to do is abandon the idea—the millennia-old idea—that pork should be cheap.

ACKNOWLEDGMENTS

The tribe of academic historians, of which I am an apostate member, would describe this book as a work of synthesis, which means I didn't root about in the archives myself but rather let others—historians, archaeologists, geographers, theologians, and literary critics—do the real work and then swooped in to reap the fruits of their labor. That characterization is not unfair. The least I can do is express my immense gratitude to those scholars, whose work I cite in the endnotes; I pray I have not excessively distorted their arguments in the interest of concision.

Matthew Baldwin, Tom Hatley, Patricia Rucidlo, and C. A. Carlson read the manuscript and offered advice, some of which, perhaps unwisely, I have ignored. For tolerating my fumbling attempts at animal husbandry during early-morning pig-chore shifts at Warren Wilson College, I offer my apologies to the pigs and my thanks to farm manager Chase Hubbard, assistant manager Jed Brown, and precocious undergraduate farmhands too numerous to mention. Casey McKissick, farmer, butcher, and former director of a state-sponsored program called NC Choices, guided me through the world of local meat. Jamie Ager, who with his wife, Amy Ager, owns Hickory Nut Gap Farm, talked to me about livestock and showed me through the house where he grew up, a nineteenth-century inn that once hosted hog drovers.

Professor William Flowers shared his deep porcine knowledge and gave me a detailed tour of the swine research facility at North Carolina State University.

I test-drove some of the ideas in this book before audiences at Zingerman's Camp Bacon in Ann Arbor, Michigan, and at the Southern Foodways Alliance (SFA) Symposium in Oxford, Mississippi; my thanks to Zingerman's cofounder Ari Weinzweig and SFA director John T. Edge for feeding me well and introducing me to many talented farmers, chefs, writers, and eaters. Jane Adkisson of Buncombe County Public Library served up the speediest interlibrary loan service known to humankind, a godsend for a researcher sans research library. Howard Yoon, friend and agent, and his partner, Gail Ross, led me through the thickets of writing and selling a book proposal. Alex Littlefield, my editor at Basic Books, inspired me with early enthusiasm and deftly poked and prodded the book into its final form. Elizabeth Dana skillfully performed the thankless tasks attendant upon her role as editorial assistant; I thank her, just as I thank many others at Basic, unknown to me as I write this, who will do their work after mine is done.

Dorothy F. Essig, World's Greatest Mom, shared some not entirely fond childhood memories of hog butchering in her hometown of Rich Fountain, Missouri, where—as tends to be true wherever people kill pigs—cleaning intestines was women's work. Though she eats no pork, Melissa Cole Essig loved this project from the get-go and loved me all the way to—and far beyond, I profoundly hope—its completion. Our children, Jack and Lydia, issued forth a steady gush of joy and displayed alarmingly large appetites for the pork shoulders I smoked and the city hams I cured (all from well-raised pigs, of course). With full mouths and greasy chins they repeatedly asked me, "If you like pigs so much, why do you eat them?" and were not much impressed with my answers.

NOTES

PROLOGUE

1 **Cowardin too cursed the pigs at first:** James Cowardin, "Letter," *Asheville Citizen,* December 5, 1878.

3 **Apparently not many people:** Caroline Grigson, "Culture, Ecology, and Pigs from the 5th to the 3rd Millennium BC Around the Fertile Crescent," in *Pigs and Humans,* ed. Umberto Albarella (New York: Oxford University Press, 2007), 100.

5 **Pigs "were generally recognized":** George Orwell, *Animal Farm* (Boston: Mariner, 2009), 32, 191.

5 **A pig that knows where food is cached:** Suzanne Held et al., "Social Tactics of Pigs in a Competitive Foraging Task," *Animal Behaviour* 59 (2000): 569–576.

5 **Animal scientist Temple Grandin reports:** Temple Grandin and Catherine Johnson, *Animals in Translation* (New York: Scribner, 2005), 99–100.

5 **Pliny the Elder claimed that pigs:** Pliny the Elder, *Natural History,* trans. John Bostock and H. T. Riley (London: George Bell & Sons, 1890), 2:343.

5 **Sows have been trained to hunt truffles:** Lyall Watson, *The Whole Hog* (London: Profile Books, 2004), 57, 173.

5 **In early nineteenth-century England, a black sow:** William Youatt and William Martin, *The Hog* (New York: A. O. Moore, 1858), 35–36.

5 **"Pigs are a race unjustly calumniated":** James Boswell, *The Life of Samuel Johnson,* ed. Alexander Napier (London: George Bell, 1889), 4:284.

6 **The trainer, dressed as a butcher:** Ricky Jay, *Learned Pigs and Fireproof Women* (New York: Villard, 1986), 27.

6 **The first pig to play Arnold Ziffel:** William Hedgepeth, *The Hog Book* (Athens: University of Georgia Press, 1998), 111.

6 **In one convenient package, it provides:** H. L. Abrams, "The Preference for Animal Protein and Fat: A Cross-Cultural Survey," in *Food and Evolution,* ed. Marvin Harris and Eric Ross (Philadelphia: Temple University Press, 1987), 207–223.

7 **"Those who could, gorged themselves":** Eugen Weber, *A Modern History of Europe* (New York: Norton, 1971), 202.

8 **Efforts to control them—including shooting:** Mark Essig, "High Above the Hog," *New York Times,* August 31, 2011.

8 **In 1699 a French scholar estimated:** Jean-Jacques Hémardinquer, "The Family Pig of the Ancien Régime," in *Food and Drink in History*, ed. Robert Forster and Orest Ranum (Baltimore: Johns Hopkins University Press, 1979), 59n1.

8 **The two-line poem "Bacon & Eggs":** The poem is widely attributed to Howard Nemerov (see, for instance, Kevin Young, *The Hungry Ear* [New York: Bloomsbury, 2012], 151), but his authorship could not be established with certainty. Author's e-mail correspondence with Alexander Nemerov, October and November 2014.

8 **The pig's certain doom has launched:** Chris Noonan, dir., *Babe* (Universal City, CA: Universal Pictures, 1995); Dick King-Smith, *The Sheep-Pig* (New York: Puffin, 1983); E. B. White, *Charlotte's Web* (New York: Harper, 1952).

9 **Homer asks her, "Are you saying":** Mark Kirkland, dir., "Lisa the Vegetarian," *The Simpsons*, Fox, aired October 15, 1995.

9 **"There is no animal that affords a greater variety":** Pliny the Elder, *Natural History*, 2:344.

9 **"Pigs! Pigs! Pork! Pork! Pork!":** Sam Bowers Hilliard, *Hog Meat and Hoecake* (Carbondale: Southern Illinois University Press, 1972), 92–93.

10 **Pigs stood at the center of cultural life:** *The Importance of the Pig in Pacific Island Culture: An Annotated Bibliography* (New Caledonia: Secretariat of the Pacific Community, 2007); Roy Rappaport, *Pigs for the Ancestors* (New Haven, CT: Yale University Press, 1968).

10 **The Chinese character for "home":** C. A. S. Williams, *Outlines of Chinese Symbolism* (Shanghai: Kelly and Walsh, 1941), 326.

10 **In a classic work of American agricultural history:** Allen G. Bogue, *From Prairie to Corn Belt* (Chicago: University of Chicago Press, 1963), 103.

11 **we use food to stigmatize foreigners:** Stephen Mennell, *All Manners of Food* (New York: B. Blackwell, 1985), 17; Emiko Ohnuki-Tierney, *Rice as Self* (Princeton, NJ: Princeton University Press, 1993), 116–118.

11 **In 2012 an Oregon farmer went to feed his sows:** Jack Moran, "Death of Farmer Eaten by Hogs Investigated," *Register-Guard* (Eugene, Oregon), October 2, 2012.

11 **an age-old expression found in many languages:** Fernand Braudel, *Capitalism and Material Life, 1400–1800* (New York: Harper and Row, 1973), 66.

13 **"There's always a certain tension about a bunch of pigs":** Hedgepeth, *Hog Book*, 53.

13 **Alice, during her adventures in Wonderland:** Lewis Carroll, *Alice's Adventures in Wonderland* (Boston: Lothrop, 1898), 52.

13 **In *The Odyssey* the enchantress Circe transforms sailors:** *Homer's Odyssey*, trans. Alexander Pope (Edinburgh: John Ross, 1870), 157.

13 **Renaissance physicians claimed that human flesh tasted:** Gananath Obeyesekere, *Cannibal Talk* (Berkeley: University of California Press, 2005), 28.

14 **Scientists are developing genetically modified pigs:** Julie Steenhuysen, "Genome Scientist Craig Venter in Deal to Make Humanized Pig Organs," Reuters, May 6, 2014.

14 **Pigs get ulcers:** Alison Abbott, "Pig Geneticists Go the Whole Hog," *Nature* 491 (2012): 315–316.

14 **"Dogs look up to you":** James C. Humes, ed., *The Wit and Wisdom of Winston Churchill* (New York: HarperCollins, 1994), 6.

CHAPTER 1

15 The tooth, Cook told Osborn: Stephen Jay Gould, "An Essay on a Pig Roast," in *Bully for Brontosaurus* (New York: Penguin, 1992), 434.

16 The *New York Times* explained that the tooth provided: "Nebraska's 'Ape Man of the Western World,'" *New York Times*, September 17, 1922.

16 Osborn did not let the irony pass unnoted: Gould, "An Essay," 436.

17 When the dinosaurs died, mammals rose: T. S. Kemp, *The Origin and Evolution of Mammals* (New York: Oxford University Press, 2005).

19 chewith the cud: Leviticus 11:3, King James Version (hereafter "KJV").

20 Since good hearing was an advantage: Donald Broom and Andrew Fraser, *Domestic Animal Behaviour and Welfare* (Wallingford, UK: CABI, 2007), 98.

21 Other muscles clamp the nostrils shut: Lyall Watson, *The Whole Hog* (London: Profile Books, 2004), 40–41.

21 Despite constant rough use, the snout: William Hedgepeth, *The Hog Book* (Athens: University of Georgia Press, 1998), 133.

21 art-gum eraser tender: Tom Hatley and John Kappelman, "Bears, Pigs, and Plio-Pleistocene Hominids," *Human Ecology* 8 (1980): 371–387.

22 One scientist who studies pig cognition complained: Natalie Angier, "Pigs Prove to Be Smart, if Not Vain," *New York Times*, November 9, 2009.

23 We might think of the pig: Watson, *Whole Hog*, 32–33; Michael Pollan, *The Omnivore's Dilemma* (New York: Penguin, 2006), 3–5.

23 By cooking their meats and roots: Richard Wrangham, *Catching Fire* (New York: Basic Books, 2009), 109–120.

24 Pigs, with simple guts and calorie-intensive diets: Leslie C. Aiello, Peter Wheeler, and David Chivers, "The Expensive-Tissue Hypothesis," *Current Anthropology* 36 (1995): 199–221.

25 The Eurasian wild boar: Martien a. M. Groenen et al., "Analyses of Pig Genomes Provide Insight into Porcine Demography and Evolution," *Nature* 491 (2012): 393–398.

CHAPTER 2

28 The fact that they did not: Michael Rosenberg et al., "Hallan Cemi Tepesi," *Anatolica* 21 (1995): 3–12; Michael Rosenberg et al., "Hallan Cemi, Pig Husbandry, and Post-Pleistocene Adaptations along the Taurus-Zagros Arc (Turkey)," *Paléorient* 24 (1998): 25–41; R. W. Redding, "Ancestral Pigs," in *Ancestors for the Pigs*, ed. Sarah Nelson (Philadelphia: University of Pennsylvania Press, 1998), 65–76.

31 The people became farmers: Graeme Barker, *The Agricultural Revolution in Prehistory* (New York: Oxford University Press, 2009); Peter Bellwood, *The First Farmers* (Oxford: Blackwell, 2004).

32 That's why one scholar has labeled agriculture: Jared Diamond, "The Worst Mistake in the History of the Human Race," *Discover Magazine*, May 1987, 64–66.

33 Throughout history people have tamed: James Serpell, *In the Company of Animals* (New York: Blackwell, 1986), 61.

34 **Perhaps most importantly, they live in groups:** Jared Diamond, *Guns, Germs, and Steel* (New York: W. W. Norton, 1998), 157–175; E. O. Price, "Behavioral Aspects of Animal Domestication," *Quarterly Review of Biology* 59 (1984): 1–32; Nerissa Russell, "The Wild Side of Animal Domestication," *Society and Animals* 10 (2002): 285–302.

34 **After that time, very few male goats lived:** M. A. Zeder, "A Critical Assessment of Markers of Initial Domestication in Goats," in *Documenting Domestication*, ed. M. A. Zeder (Berkeley: University of California Press, 2006), 202–205.

34 **It's also likely that pigs were domesticated:** U. Albarella et al., "The Domestication of the Pig," in Zeder, *Documenting Domestication*, 209–227.

36 **Dogs were the first domestic animals:** Robert K. Wayne and Elaine A. Ostrander, "Lessons Learned from the Dog Genome," *Trends in Genetics* 23 (2007): 557–567; Z.-L. Ding et al., "Origins of Domestic Dog in Southern East Asia Is Supported by Analysis of Y-Chromosome DNA," *Heredity* 108 (2011): 507–514.

36 **Modern experiments show that wolves hand-raised:** Raymond Coppinger and Lorna Coppinger, *Dogs* (New York: Simon and Schuster, 2001), 39–50.

37 **A genetic mutation that allowed wolves:** E. Axelsson et al., "The Genomic Signature of Dog Domestication Reveals Adaptation to a Starch-Rich Diet," *Nature* 495 (2013): 360–364.

37 **Some of the waste would have been cooked:** Richard Wrangham, *Catching Fire* (New York: Basic Books, 2009), 55–81.

37 **The wild animals began to separate:** Carlos Driscoll et al., "From Wild Animals to Domestic Pets," *Proceedings of the National Academy of Sciences* 106 (2009): 9971–9978.

38 **The boars and wolves most adept:** Coppinger and Coppinger, *Dogs*, 50–67; Albarella et al., "Domestication of the Pig."

38 **And over that span, the pig bones change:** Anton Ervynck et al., "Born Free? New Evidence for the Status of *Sus scrofa* at Neolithic Cayonu Tepesi," *Paléorient* 27 (2001): 47–73; Hitomi Hongo and Richard H. Meadow, "Pig Exploitation at Neolithic Cayonu Tepesi," in Nelson, *Ancestors for the Pigs*, 77–98; J. Conolly et al., "Meta-Analysis of Zooarchaeological Data from SW Asia and SE Europe Provides Insight into the Origins and Spread of Animal Husbandry," *Journal of Archaeological Science* 38 (2011): 538–545.

CHAPTER 3

43 **Whatever the technical method, building the pyramids:** Barry Kemp, *Ancient Egypt* (New York: Routledge, 1989).

44 **This proved that Giza was a provisioned site:** Richard Redding, "Status and Diet at the Workers' Town, Giza, Egypt," in *Anthropological Approaches to Zooarchaeology*, ed. D. Campana et al. (Oxford: Oxbow Books, 2010); Mark Lehner, "Villages and the Old Kingdom," in *Egyptian Archaeology*, ed. Willeke Wendrich (Malden, MA: Wiley-Blackwell, 2010), 85–101.

44 **Villagers at Kom el-Hisn raised cattle:** Richard Redding, "Egyptian Old Kingdom Patterns of Animal Use and the Value of Faunal Data in Modeling Socioeconomic Systems," *Paléorient* 18 (1992): 99–107; Richard Redding, "The Role of the Pig in the Subsistence System of Ancient Egypt," in *Ancestors for the Pigs*, ed. Sarah Nelson (Philadelphia: University of Pennsylvania Press, 1998), 20–30.

46 **"A human being is primarily a bag":** George Orwell, *The Road to Wigan Pier* (New York: Harcourt Brace Jovanovich, 1958), 91.

46 **The records make no mention of pigs:** M. A. Zeder, "Of Kings and Shepherds," in *Chiefdoms and Early States in the Near East*, ed. Gil Stein and Mitchell S. Rothman (Madison, WI: Prehistory Press, 1994), 175–191.

46 **In other words, if it was biologically possible:** Caroline Grigson, "Culture, Ecology, and Pigs from the 5th to the 3rd Millennium BC Around the Fertile Crescent," in *Pigs and Humans*, ed. Umberto Albarella (New York: Oxford University Press, 2007).

47 **That's when the villagers turned to pigs:** Brian Hesse, "Pig Lovers and Pig Haters," *Journal of Ethnobiology* 10 (1990): 195–225; Brian Hesse and Paula Wapnish, "Can Pig Remains Be Used for Ethnic Diagnosis in the Ancient Near East?," in *Archaeology of Israel*, ed. N. A. Silberman and D. Small (Sheffield, UK: Sheffield Academic Press, 1997), 238–270; M. A. Zeder, *Feeding Cities* (Washington, DC: Smithsonian Institution Press, 1991).

47 **As pigs lost habitat:** Marvin Harris, *The Sacred Cow and the Abominable Pig* (New York: Simon & Schuster, 1987), 75–77.

47 **A thousand years later, few people:** Hesse, "Pig Lovers," 218.

48 **Archaeologists tend to find pig bones:** M. A. Zeder, "Pigs and Emergence Complexity in the Ancient Near East," *Masca Research Papers in Science and Archaeology* 15 (1998): 118; Joanna Piątkowska-Małecka and Anna Smogorzewska, "Animal Economy at Tell Arbid, Northeast Syria, in the Third Millennium BC," *Bioarchaeology of the Near East* 4 (2010): 25–43; K. Mudar, "Early Dynastic III Animal Utilization in Lagash," *Journal of Near Eastern Studies* 41 (1982): 23–34; H. M. Hecker, "A Zooarchaeological Inquiry into Pork Consumption in Egypt from Prehistoric to New Kingdom Times," *Journal of the American Research Center in Egypt* 19 (1982): 59–71.

48 **Although absent from the residences of official workers:** Redding, "Status and Diet."

48 **The Greek historian Herodotus, in the fifth century BC:** Herodotus, *The History*, trans. David Grene (Chicago: University of Chicago Press, 1987), 151.

49 **Residents threw garbage into the streets:** Elizabeth Stone, "The Spatial Organization of Mesopotamian Cities," *Aula Orientalis* 9 (1991): 235–242.

49 **"You shall have a stick":** Deuteronomy 23:12–14, Revised Standard Version (hereafter "RSV").

49 **A few elite homes and temples had pit latrines:** Marc Van de Mieroop, *The Ancient Mesopotamian City* (New York: Oxford University Press, 1999), 159–160.

49 **In many villages around the world today:** D. W. Gade, "The Iberian Pig in the Central Andes," *Journal of Cultural Geography* 7 (1987): 35–49.

49 **Some English pigs in the eighteenth and nineteenth centuries:** Robert Malcolmson, *The English Pig* (London: Hambledon, 2001), 5–7.

49 **The structure was originally identified as a grain silo:** F. Bray, "Agriculture," in *Science and Civilization in China*, ed. J. Needham (Cambridge: Cambridge University Press, 1984), 6:291–292.

50 **The practice was widespread:** E. Anderson, *The Food of China* (New Haven, CT: Yale University Press, 1988), 125.

50 **In the 1960s more than 90 percent of farmers:** D. J. Nemeth, "Privy-Pigs in Prehistory?," in Nelson, *Ancestors for the Pigs*, 16.

50 **Since these eggs are produced:** Robert L. Miller, "Hogs and Hygiene," *Journal of Egyptian Archaeology* 76 (1990): 130.

50 **In Aristophanes' play *Peace:*** Aristophanes, *Peace*, in *Eleven Comedies* (New York: Tudor, 1934), 154.

51 **Eating human flesh and eating excrement:** William Miller, *The Anatomy of Disgust* (Cambridge, MA: Harvard University Press, 1997), 15, 62.

51 **"The pig is impure":** JoAnn Scurlock, "Animal Sacrifice in Ancient Mesopotamian Religion," in *A History of the Animal World in the Ancient Near East*, ed. Billie Jean Collins (Boston: Brill, 2002), 393.

51 **"May dogs and swine eat your flesh":** Walter Houston, *Purity and Monotheism* (Sheffield: JSOT Press, 1993), 190.

51 **The people of the Near East practiced:** Edwin Firmage, "Zoology," in *Anchor Bible Dictionary*, ed. D. N. Freedman (New York: Doubleday, 1992), 6:1109–1167.

51 **In Mesopotamia and Egypt, pigs never:** J. N. Postgate, *Early Mesopotamia* (London: Routledge, 1992), 166; Douglas Brewer, "Hunting, Animal Husbandry and Diet in Ancient Egypt," in Collins, *History of the Animal World in the Ancient Near East*, 440–443.

51 **Pork does not appear on the list:** William J. Darby, *Food, the Gift of Osiris* (New York: Academic Press, 1977), 175.

51 **"The pig is not fit for a temple":** Scurlock, "Animal Sacrifice," 393.

51 **If anyone served the gods:** Billie Jean Collins, "Pigs at the Gate," *Journal of Ancient Near Eastern Religions* 6 (2006): 156–157.

CHAPTER 4

53 **"I will indeed bless you":** Genesis 22:17, RSV.

54 **Among the forbidden beasts were pigs:** Leviticus 11:8, KJV.

54 **These settlers were the Israelites:** Daniel Snell, *Religions of the Ancient Near East* (New York: Cambridge University Press, 2011), 104.

54 **Israelite priests, in banning pork:** Walter Houston, *Purity and Monotheism* (Sheffield, UK: JSOT Press, 1993), 171; Brian Hesse and Paula Wapnish, "Can Pig Remains Be Used for Ethnic Diagnosis in the Ancient Near East?," in *Archaeology of Israel*, ed. N. A. Silberman and D. Small (Sheffield, UK: Sheffield Academic Press, 1997).

55 **Douglas's argument, though, suffers from circularity:** Mary Douglas, *Purity and Danger* (London: Routledge and Kegan Paul, 1966), 54–55; Mary Douglas, "Deciphering a Meal," *Daedalus* 101 (1972): 71; R. Bulmer, "Why Is the Cassowary Not a Bird?," *Man* 2 (1967): 21.

55 **The pork prohibition therefore simply codified:** Marvin Harris, *The Sacred Cow and the Abominable Pig* (New York: Simon & Schuster, 1987), 82–86.

56 **This was the case with nomadic Mongols:** Frederick Simoons, *Eat Not This Flesh* (Madison: University of Wisconsin Press, 1994), 82.

56 **When they did, the sacrifices:** Houston, *Purity and Monotheism*, 72, 149, 253.

56 **Though promising, this theory rests:** P. Diener et al., "Ecology, Evolution, and the Search for Cultural Origins," *Current* 19 (1978): 493–540.

57 **Just about any kind of meat:** Harris, *Sacred Cow*, 69–71.

57 **"God forbid that I should believe":** Simoons, *Eat Not This Flesh*, 71.

58 **The key rule was this:** Leviticus 11:3, KJV.

58 **The same rule disqualified pigs:** Leviticus 11:3, 7–8, KJV.

58 **Diet played an important role in scripture:** Genesis 1:29–30, KJV.

58 God told Noah that he could eat: Genesis 9:2–3, RSV.

58 "You shall not eat flesh with its life": Genesis 9:4, RSV.

58 which was thought to contain the "life force": Jonathan Brumberg-Kraus, "Meat-Eating and Jewish Identity," *AJS Review* 24 (1999): 241–242.

58 "Eat not the blood": Deuteronomy 12:23, KJV.

59 Deuteronomy forbids eating carrion: Deuteronomy 14:21, RSV.

59 In the Christian Bible Jesus advises: Matthew 7:6, KJV.

59 "As a dog returneth to his vomit": Proverbs 26:11, KJV.

59 According to the book of Kings, "Thus says the Lord": 1 Kings 21:19, 22:37–38, KJV.

59 But in the Septuagint, a Greek translation of the Jewish Bible: Houston, *Purity and Monotheism*, 190–191.

60 Uncleanliness, in the Bible, is a contagion: Houston, *Purity and Monotheism*, 145–146; Mary Douglas, "The Forbidden Animals in Leviticus," *Journal for the Study of the Old Testament* 8 (1993): 3–23; Calum Carmichael, "On Separating Life and Death: An Explanation of Some Biblical Laws," *Harvard Theological Review* 69 (1976): 1–7.

60 Then, starting in about 300 BC: Brian Hesse, "Pig Lovers and Pig Haters," *Journal of Ethnobiology* 10 (1990); Hesse and Wapnish, "Can Pig Remains"; Jordan Rosenblum, "'Why Do You Refuse to Eat Pork?' Jews, Food, and Identity in Roman Palestine," *Jewish Quarterly Review* 100 (2009): 96–97.

60 Many Jews acquiesced: 1 Maccabees 1: 41–43, RSV.

60 Worst of all, Antiochus ordered the Jews: 1 Maccabees 1: 46–48, RSV.

61 His purpose, he explains, is to leave: 2 Maccabees 6:18–31, RSV.

61 After he is dead, they kill another: 2 Maccabees 7:1–41, RSV; Molly Whittaker, *Jews and Christians* (New York: Cambridge University Press, 1984), 73.

61 It is a matter between the Lord and his people: Isaiah 65:3–4, 66:17, KJV.

63 Now, however, it also became a way: It has been argued that the Israelites abstained from pork to distinguish themselves from the Philistines 1,000 years earlier, but the evidence for this is uncertain. See Hesse and Wapnish, "Can Pig Remains," 248.

63 Jews "do not differentiate": Whittaker, *Jews and Christians*, 76

63 It was said that Caesar Augustus: Rosenblum, "Why Do You Refuse," 99.

CHAPTER 5

65 "Thrushes, fatted hens, bird gizzards!": Federico Fellini, dir., *Satyricon* (Produzioni Europee Associati, 1969) (quotation from English subtitles).

66 Petronius also describes a whole roast pig: Petronius, *Satyricon*, trans. Alfred Allinson (Paris: Charles Carrington, 1902), 110.

66 "I declare my cook made it": Petronius, *Satyricon*, 190.

66 In cuisine, culture, and mythology, Romans delighted: Mireille Corbier, "The Ambiguous Status of Meat in Ancient Rome," *Food and Foodways* 3 (1989): 240–241, 248.

67 In Greek mythology, after Jason and Medea kill: Apollonius Rhodius, *Argonautica*, trans. R. C. Seaton (New York: MacMillan, 1912), 343.

67 Similarly, a painted vase shows Apollo: Judith Yarnall, *Transformations of Circe* (Urbana: University of Illinois Press, 1994), 46.

67 **Romans killed pigs to seal public agreements:** Daniel Ogden, *A Companion to Greek Religion* (Oxford: Blackwell, 2007), 133.

67 **The rotted pork was then scattered:** Walter Burkert, *Greek Religion* (Cambridge, MA: Harvard University Press, 1985), 13, 242–244.

67 **In Greece young pigs were known by the terms:** Andrew Dalby, *Food in the Ancient World from A to Z* (London: Psychology Press, 2003), 269.

67 **Aristophanes makes some horrifying puns:** Aristophanes, *Aristophanes*, ed. David R. Slavitt and Palmer Bovie (Philadelphia: University of Pennsylvania Press, 1998), 1:45–46.

67 **The scholar Varro noted that Romans:** Marcus Terentius Varro, *On Agriculture*, trans. William Davis Hooper and Harrison Boyd Ash (Cambridge, MA: Harvard University Press, 1979), 357.

68 **Sacrificing pigs honored the gods:** Marija Gimbutas, *The Goddesses and Gods of Old Europe* (Berkeley: University of California Press, 1982), 214–215.

68 **This was the cheapest way:** Peter Garnsey, *Food and Society in Classical Antiquity* (New York: Cambridge University Press, 1999), 13–17, 123.

69 **There were more Latin words for pork:** H. J. Loane, *Industry and Commerce of the City of Rome* (Philadelphia: Arno Press, 1979), 127.

69 **According to the Edict of Diocletian:** Michael MacKinnon, *Production and Consumption of Animals in Roman Italy* (Portsmouth, RI: Journal of Roman Archaeology, 2004), 208–209.

69 **After the Punic Wars, the percentage of pig bones:** Michael MacKinnon, "'Romanizing' Ancient Carthage," in *Anthropological Approaches to Zooarchaeology*, ed. Douglas Campana et al. (Oxford: David Brown, 2010), 172.

69 **Other sections of the book offer recipes:** Christopher Grocock and Sally Grainger, *Apicius* (Devon, UK: Prospect Books, 2006), 55–56, 70.

69 **Archeology confirms that Romans carved up pigs:** MacKinnon, *Production and Consumption*, 168.

69 **Apicius is credited with inventing the technique:** Pliny the Elder, *Natural History*, trans. John Bostock and H. T. Riley (London: George Bell & Sons, 1890), 2:344.

70 **Finally, the stomach is tied:** Grocock and Grainger, *Apicius*, 247–249.

70 **The Roman poet Martial had this to say:** Martial, *Epigrams* (London: Bell & Daldy, 1865), 593.

70 **Elsewhere, after a meal, Martial suffers the glutton's regret:** Martial, *Epigrams*, 313.

71 **The womb of this poor sow:** Plutarch, *Moralia*, trans. Harold F. Cherniss (Cambridge, MA: Harvard University Press, 1957), 12:565.

71 **Seneca, the Stoic philosopher and statesman, decried:** Corbier, "Ambiguous Status," 241.

71 **By 450 AD about 140,000 citizens:** S. J. B. Barnish, "Pigs, Plebeians and Potentes," *Papers of the British School at Rome* 55 (1987): 160–165; A. H. M. Jones, *The Later Roman Empire, 284–602* (Baltimore: Johns Hopkins University Press, 1964), ii, 696.

72 **By contrast, imports from outside the Italian Peninsula:** J. Hughes, *Pan's Travail* (Baltimore: Johns Hopkins University Press, 1994), 146.

72 **Grain sufficient to feed hundreds of thousands of people:** Peter Temin, *The Roman Market Economy* (Princeton, NJ: Princeton University Press, 2013), 29–31.

72 **Beef and mutton came from older animals:** Michael Ross MacKinnon, "Animal Production and Consumption in Roman Italy" (PhD diss., University of Alberta, 1999), 78–80, 97–98, 112–113, 209–210, 237.

72 **A popular saying held:** Pliny the Elder, *Natural History*, 2:343.

73 **According to Varro, Rome's most important agricultural writer:** Varro, *On Agriculture*, 357.

73 **Varro devoted more attention to pigs than to cows:** Varro, *On Agriculture*, 353.

73 **Columella, writing in the first century** AD, **extolled:** Lucius Junius Moderatus Columella, *On Agriculture*, trans. Harrison Boyd Ash (Cambridge, MA: Harvard University Press, 1960), 2:293.

73 **Boars, Columella tells us, should possess:** Columella, *On Agriculture*, 291.

74 **With that sort of production, farmers had the incentive:** Varro, *On Agriculture*, 365.

74 **The smaller looked like a downsized wild boar:** Columella, *On Agriculture*, 291.

74 **The best feeding grounds for such pigs:** Columella, *On Agriculture*, 293.

74 **Columella also described the larger variety:** Columella, *On Agriculture*, 291.

74 **Varro reports that nursing sows were fed:** Varro, *On Agriculture*, 361.

74 **These fat white pigs were kept closer to Rome:** Petronius, *Satyricon*, 131.

74 **Columella advises that on all farms:** Columella, *On Agriculture*, 291.

75 **And they made impressive offerings:** Michael MacKinnon, "High on the Hog: Linking Zooarchaeological, Literary, and Artistic Data for Pig Breeds in Roman Italy," *American Journal of Archaeology* 105 (2001): 667.

75 **A pig that sups on fish guts:** Richard Bradley, *Gentleman and Farmers Guide* (London: W. Mears, 1732), 71.

CHAPTER 6

78 **Rome's complex networks of Mediterranean commerce:** Bryan Ward-Perkins, *The Fall of Rome* (New York: Oxford University Press, 2005).

78 **Archaeologists digging in post-Roman sites:** S. White, "From Globalized Pig Breeds to Capitalist Pigs," *Environmental History* 16 (2011): 100.

79 **Another group moved overland out of Turkey and Greece:** Peter Rowley-Conwy, "Westward Ho! The Spread of Agriculture from Central Europe to the Atlantic," *Current Anthropology* 52, suppl. 4 (2011): S431–S451.

79 **Genetic studies tell us that the first wave:** Greger Larson et al., "Phylogeny and Ancient DNA of *Sus*," *Proceedings of the National Academy of Sciences* 104 (2007): 4834–4839.

80 **An Irish myth tells of pigs:** Miranda Green, *Dictionary of Celtic Myth and Legend* (New York: Thames and Hudson, 1992), 44–45; Jeffrey Greene, *The Golden-Bristled Boar* (Charlottesville: University of Virginia Press, 2011).

80 **From these laws we can infer:** Katherine Fischer Drew, *The Laws of the Salian Franks* (Philadelphia: University of Pennsylvania Press, 1991), 7, 3–5, 66–73, 88.

80 **Anglo-Saxons valued a pig:** Joyce Salisbury, *The Beast Within* (New York: Routledge, 1994), 34.

81 **In a practice known as *denbera*:** John Thrupp, "On the Domestication of Certain Animals in England Between the Seventh and Eleventh Centuries," *Transactions of the Ethnological Society of London* 4 (1866): 164–172.

81 **In England's Domesday Book:** Robert Trow-Smith, *A History of British Livestock Husbandry to 1700* (London: Routledge and Kegan Paul, 1957), 51.

81 **In ninth-century Italy a monastery's forest:** Vito Fumagalli, *Landscapes of Fear*, trans. Shayne Mitchell (Cambridge, MA: Blackwell, 1994), 146.

81 **The tips of stone arrowheads have been found:** U. Albarella and D. Serjeantson, "A Passion for Pork," in *Consuming Passions and Patterns of Consumption*, ed. P. Miracle and N. Milner (Cambridge, UK: Monographs of the McDonald Institute), 44.

81 **"It is dangerous for one unfamiliar with their ways to approach them":** Strabo, *Geography*, trans. Horace Leonard Jones (New York: G. P. Putnam's Sons, 1923), 2:243.

81 **In the forests of Kent in the ninth century:** Caroline Grigson, "Porridge and Pannage," in *Archaeological Aspects of Woodland Ecology*, ed. Martin Bell and Susan Limbrey (Oxford: British Archaeological Reports, 1982), 300–301.

82 **Swineherds carried either a long, slender pole:** Earl Shaw, "Geography of Mast Feeding," *Economic Geography* 16 (1940): 233–249.

82 **An English law of 1184 decreed:** Robert Bartlett, *England Under the Norman and Angevin Kings* (New York: Oxford University Press, 2000), 239, 674.

83 **Lions and leopards kill with claws and teeth:** John Cummins, *Hound and Hawk* (New York: St. Martin's Press, 1988), 96.

83 **The boar slashes at an approaching hero:** Ovid, *Metamorphoses*, trans. Henry T. Riley (London: George Bell, 1893), 279, 281.

84 **Arthur became known as the Boar of Cornwall:** *The Mabinogion* (London: Quaritch, 1877), 239–257.

84 **In present-day Belgium, bones dug up at castles:** Anton Ervynck, "Orant, Pugnant, Laborant," in *Behaviour Behind Bones*, ed. W. Van Neer and A. Ervynck (Oxford: Oxbow, 2004), 215–223.

84 **The trash heaps of the elite:** Annie Grant, "Food, Status and Social Hierarchy," in Miracle and Milner, *Consuming Passions*, 18; R. M. Thomas, "Food and the Maintenance of Social Boundaries in Medieval England," in *Archaeology of Food and Identity*, ed. K. C. Twiss (Carbondale, IL: Center for Archaeological Investigations, 2007), 138–144.

84 **Medieval Europeans ate spices because they liked them:** Paul Freedman, *Out of the East* (New Haven, CT: Yale University Press, 2008), 3–6, 19–25.

84 **Medieval cooks also borrowed from Rome:** Bridget Ann Henisch, *Fast and Feast* (University Park: Pennsylvania State University Press, 1976), 131.

85 **A camphor-soaked wick was placed in the boar's mouth:** Freedman, *Out of the East*, 37.

85 **One cookbook offered a recipe for a roasted rooster:** Freedman, *Out of the East*, 38.

85 **In noble houses, the pantry of preserved foods:** Terence Scully, *The Art of Cookery in the Middle Ages* (Woodbridge, UK: Boydell Press, 1995), 244.

85 **Sometimes the salt gets an assist:** R. Lawrie, *Lawrie's Meat Science* (Boca Raton, FL: Woodhead Publishing, 2006), 130–132.

86 **Greeks used the same word to describe:** Frank Frost, "Sausage and Meat Preservation in Antiquity," *Greek, Roman, and Byzantine Studies* 40 (1999): 244.

86 **According to Cato, "No moths nor worms will touch":** Marcus Cato, *On Agriculture*, trans. William Davis Hooper and Harrison Boyd Ash (Cambridge, MA: Harvard University Press, 1935), 154–157.

86 **Varro insisted that the Gauls:** Marcus Terentius Varro, *On Agriculture*, trans. William Davis Hooper and Harrison Boyd Ash (Cambridge, MA: Harvard University Press, 1979), 357.

86 **These Gauls lived around Parma:** David Thurmond, *Handbook of Food Processing in Classical Rome* (Boston: Brill, 2006), 217.

86 **Varro recommended pork from what is now Portugal:** Strabo, *Geography*, 101.

86 **Martial gave a nod to hams:** Martial, *Epigrams* (London: Bell & Daldy, 1865), 595.

86 **Living things need to eat fats:** Harold McGee, *On Food and Cooking*, rev. ed. (New York: Scribner, 2004), 797; Adam Drewnowski, "Why Do We Like Fat?," *Journal of the American Dietetic Association* 97, suppl. (1997): S58–S62.

87 **Ancient cooks often boiled their meat:** Mireille Corbier, "The Ambiguous Status of Meat in Ancient Rome," *Food and Foodways* 3 (1989): 233.

87 **Fat was so rare and precious:** Harry Hoffner, "Oil in Hittite Texts," *The Biblical Archaeologist* 58 (2012): 109.

87 **a sort of olive tree on the hoof:** Ari Weinzweig, *Zingerman's Guide to Better Bacon* (Ann Arbor, MI: Zingerman's Press, 2009), 30–38.

87 **For medieval Europeans, the seasons were a bumpy cycle:** Bartlett, *England Under*, 643; Dolly Jørgensen, "Pigs and Pollards," *Sustainability* 5 (2013): 387–399.

87 **Many proverbs indicated that a supply:** Jean-Jacques Hémardinquer, "The Family Pig of the Ancien Régime," in *Food and Drink in History*, ed. Robert Forster and Orest Ranum (Baltimore: Johns Hopkins University Press, 1979), 58.

CHAPTER 7

90 **The patron saint of animals expressed no sympathy:** Thomas Okey Francis and Robert Steele, *The Little Flowers of St. Francis* (New York: E. P. Dutton & Co., 1910), 134–137.

90 **"Cursed be that evil beast":** Saint Bonaventure, *Life of Saint Francis* (London: J. M. Dent, 1904), 85–86.

91 **According to the New Testament, Christ was:** John 1:29, RSV.

91 **The Christian Bible picked up this theme:** Psalms 23:1 KJV.

91 **According to the Second Epistle of Peter:** 2 Peter 2:22, RSV.

91 **The prodigal son, after squandering his inheritance:** Luke 15:16, KJV.

91 **Those husks, incidentally, were likely pods:** John Russell Smith, *Tree Crops* (New York: Harcourt, 1929), 33–34.

92 **He said to the demons, "Go":** Matthew 8:32, KJV.

92 **In the words of one bestiary:** Richard Barber, *Bestiary* (Woodbridge, UK: Boydell, 1993), 81.

92 **"The pig is a filthy beast":** Barber, *Bestiary*, 84–86.

92 **In addition to filth, the pig stood:** Albertus Magnus, *Questions Concerning Aristotle's "On Animals,"* ed. Irven Resnick (Washington, DC: Catholic University of America Press, 2008), 239.

93 **During sex the boar's penis:** Wilson G. Pond and Harry J. Mersmann, *Biology of the Domestic Pig* (Ithaca, NY: Cornell University Press, 2001).

93 **An early agricultural writer described pigs:** Gervase Markham, *Cheape and Good Husbandry* (London: Thomas Snodham, 1614), 100.

93 **Peter protests that he has "never eaten":** Acts 10:10–15, KJV.

93 **At one of the councils of Antioch:** Claudine Fabre-Vassas, *The Singular Beast* (New York: Columbia University Press, 1997), 6, 325–326.

94 **Even the Acts of the Apostles hedged its bets:** Acts 15:29, KJV; David Freidenreich, *Foreigners and Their Food* (Berkeley: University of California Press, 2011), 94–95, 102, 133.

94 **An Irish text warned people not to eat:** David Grumett, "Mosaic Food Rules in Celtic Spirituality in Ireland," in *Eating and Believing*, ed. D. Grumett and R. Muers (New York: T & T Clark, 2008), 35.

94 **"Swine that taste the flesh or blood of men":** John McNeill, *Medieval Handbooks of Penance* (New York: Columbia University Press, 1990), 130–135; R. Meens, "Eating Animals in the Early Middle Ages," in *The Animal/Human Boundary*, ed. Angela Creager et al. (Rochester, NY: University of Rochester Press, 2002), 15.

94 **If swine merely tasted human blood:** Grumett, "Mosaic," 34.

95 **Medieval armies could be slow to collect their dead:** Philippe Ariès, *The Hour of Our Death* (New York: Knopf, 1981), 43–44; C. Smith, "A Grumphie in the Sty: An Archaeological View of Pigs in Scotland," *Proceedings of the Society of Antiquaries of Scotland* 130 (2000): 715.

95 **In Shakespeare's *Richard III*:** William Shakespeare, *Richard III*, ed. Anthony Hammond (London: Arden Shakespeare, 2006), 306 (V.iii.7–10).

95 **"Cows feed only on grass and the leaves of trees":** McNeill, *Medieval Handbooks*, 130–135.

95 **The process was nudged along:** Thomas Benjamin, *The Atlantic World* (New York: Cambridge University Press, 2009), 38–39; Edward Barbier, *Scarcity and Frontiers* (Cambridge: Cambridge University Press, 2011), 197.

95 **In response, Parisian authorities banned pigs:** E. P. Evans, *The Criminal Prosecution and Capital Punishment of Animals* (London: Faber, 1987), 158.

95 **Similarly, in 1301 the English city of York passed:** P. J. P. Goldberg, "Pigs and Prostitutes," in *Young Medieval Women*, ed. Katherine Lewis et al. (New York: St. Martin's Press, 1999), 172.

96 **The wealthier might have pit latrines:** Alain Corbin, *The Foul and the Fragrant* (Cambridge, MA: Harvard University Press, 1986), 27.

96 **A set of German playing cards from 1535:** Peter Stallybrass, *The Politics and Poetics of Transgression* (Ithaca, NY: Cornell University Press, 1986), 57.

96 **The theologian Honorius of Autun:** Caroline Bynum, *The Resurrection of the Body in Western Christianity* (New York: Columbia University Press, 1995), 148.

96 **In an English text, a woman explains:** Irven Resnick, *Marks of Distinction* (Washington, DC: Catholic University of America Press, 2012), 14.

96 **In France in 1494, for example:** Evans, *Criminal Prosecution*, 155–156.

96 **The earliest medieval animal trials date:** J. Enders, "Homicidal Pigs and the Antisemitic Imagination," *Exemplaria* 14 (2002): 206.

96 **"When an ox gores a man or a woman to death":** Exodus 21:28, RSV.

97 **Pigs accounted for well over half:** E. Cohen, "Animals in Medieval Perceptions," in *Animals and Human Society* (New York: Routledge, 1994), 74.

97 **In modern-day Papua New Guinea:** Robert L. Miller, "Hogs and Hygiene," *Journal of Egyptian Archaeology* 76 (1990): 125.

97 **One European court explained that a pig:** Evans, *Criminal Prosecution*, 155.

97 **In another case the court noted:** Evans, *Criminal Prosecution*, 156.

97 "But if another animal or a Jew do it": Enders, "Homicidal Pigs," 230.

98 English illustrations of the crucifixion: Wendelien Van Welie-Vink, "Pig Snouts as Sign of Evil in Manuscripts from the Low Countries," *Quaerendo* 26 (1996): 213–228.

98 Martin Luther, in a religious tract, addressed Jews directly: Martin Luther, *Works* (St. Louis, MO: Concordia, 1955), 47:212; Stephen Greenblatt, "Filthy Rites," *Daedalus* 111 (1982): 11–12.

98 "If swine were used for food": Resnick, *Marks of Distinction,* 169.

100 According to one authority, no other food: Ken Albala, *Eating Right in the Renaissance* (Berkeley: University of California Press, 2002), 69.

100 The twelfth-century text *Anatomia porci:* P. Beullens, "Like a Book Written by God's Finger," in *A Cultural History of Animals in the Medieval Age,* ed. Brigitte Resl (London: Bloomsbury Academic, 2009), 146.

100 One medical book reported: William Mead, *The English Medieval Feast* (New York: Barnes & Noble, 1967), 79.

100 A butcher reportedly passed off human flesh: Albala, *Eating Right,* 69.

100 Christians, one authority explained, can transform: Irven Resnick, "Dietary Laws in Medieval Christian-Jewish Polemics," *Studies in Christian-Jewish Relations* 6 (2011): 11, 15.

101 An English rhyme told the tale of Hugh of Lincoln: Fabre-Vassas, *Singular Beast,* 134.

101 These invented tales had brutally real effects: R. Hsia, *The Myth of Ritual Murder* (New Haven, CT: Yale University Press, 1988), 1–4.

101 A London town ordinance of 1419: Resnick, *Marks of Distinction,* 153.

102 According to the Quran, the word of God: Quran 5:3.

102 Environmental and political reasons—the unsuitability of swine: P. Diener et al., "Ecology, Evolution, and the Search for Cultural Origins," *Current* 19 (1978): 493–540.

102 One Christian text depicts Jews lamenting: Resnick, *Marks of Distinction,* 157.

103 Many converts tried to combat such suspicions: B. J. Zadik, "The Iberian Pig in Spain and the Americas at the Time of Columbus" (master's thesis, University of California, Berkeley, 2005), 11–14.

103 In a work by the great playwright Lope de Vega: Rebecca Earle, *The Body of the Conquistador* (New York: Cambridge University Press, 2012), 61.

103 A convert named Gonzolo Perez Jarada: Resnick, *Marks of Distinction,* 165–166.

103 Then, after cords were twisted tightly around her wrists: Cecil Roth, *History of the Marranos* (New York: Schocken Books, 1974), 110–116.

CHAPTER 8

105 When cooked and served at a "feast among the nobles": Walter Scott, *Ivanhoe* (Edinburgh: A. and C. Black, 1860), 46–47.

106 Thus "swineflesh" became pork: Ina Lipkowitz, *Words to Eat By* (New York: St. Martin's Press, 2011), 179–183.

107 This prompted farmers to clear forests: Thomas Benjamin, *The Atlantic World* (New York: Cambridge University Press, 2009), 38–39; Edward Barbier, *Scarcity and Frontiers* (Cambridge: Cambridge University Press, 2011), 197.

107 the human population grew in tandem with the supply of grain: Fernand Braudel, *The Structures of Everyday Life* (New York: Harper and Row, 1981), 104.

107 Nobles continued to eat large amounts: Immanuel Wallerstein, *The Modern World-System* (Berkeley: University of California Press, 2011), 1:35–39.

108 In France and Germany the price of grain: P. Edwards, "Domesticated Animals in Renaissance Europe," in *A Cultural History of Animals in the Renaissance*, ed. Bruce Boehrer (Oxford: Berg, 2009), 77.

108 In 1397 the average resident of Berlin: Wilhelm Abel, *Agricultural Fluctuations in Europe* (London: Methuen, 1980), 71.

108 On one manor in Norfolk, England: Christopher Dyer, "Change in Diet in the Late Middle Ages: The Case of Harvest Workers," *Agricultural History Review* 36 (1988): 21–37.

108 In 1501 the Duke of Buckingham hosted a meal: R. M. Thomas, "Food and the Maintenance of Social Boundaries in Medieval England," in *Archaeology of Food and Identity*, ed. K. C. Twiss (Carbondale, IL: Center for Archaeological Investigations, 2007), 144–146.

109 Woodcut illustrations of peasant weddings from this era: Paul Freedman, *Out of the East* (New Haven, CT: Yale University Press, 2008), 41.

109 One Renaissance doctor advised that the sedentary elite: Ken Albala, *Eating Right in the Renaissance* (Berkeley: University of California Press, 2002), 69, 192.

109 For reasons of status, health, or both: Fernand Braudel, *Capitalism and Material Life, 1400–1800* (New York: Harper and Row, 1973), 128–132.

109 A century later another Frenchman noted: Jean-Jacques Hémardinquer, "The Family Pig of the Ancien Régime," in *Food and Drink in History*, ed. Robert Forster and Orest Ranum (Baltimore: Johns Hopkins University Press, 1979), 60n11.

109 In Scotland, another writer reported: C. Smith, "A Grumphie in the Sty: An Archaeological View of Pigs in Scotland," *Proceedings of the Society of Antiquaries of Scotland* 130 (2000): 716.

110 The poor, among their many misfortunes: The same was true in ancient Rome. See J. Toner, *Leisure and Ancient Rome* (Cambridge, MA: Polity Press, 1995), 68.

110 The English sometimes referred to a brothel: P. J. P. Goldberg, "Pigs and Prostitutes," in *Young Medieval Women*, ed. Katherine Lewis et al. (New York: St. Martin's Press, 1999), 172–173, 186n3.

110 The most common Greek word for sausage: Frank Frost, "Sausage and Meat Preservation in Antiquity," *Greek, Roman, and Byzantine Studies* 40 (1999): 246–247.

110 Shakespeare's Falstaff, a man of large and indelicate appetites: William Shakespeare, *Henry IV, Part 2*, in *The Works of William Shakespeare* (London: G. Routledge, 1869), 3:31 (II.iv.232–3).

110 Zeal-of-the-Land Busy, a Puritan intent: Peter Stallybrass, *The Politics and Poetics of Transgression* (Ithaca, NY: Cornell University Press, 1986), 63.

111 By 1696, England had about 12 million: B. H. Slicher Van Bath, "Agriculture in the Vital Revolution," *Cambridge Economic History of Europe* 5 (1977): 89.

111 Gervase Markham, in a 1614 book: Gervase Markham, *Cheape and Good Husbandry* (London: Thomas Snodham, 1614), 99–100.

111 In *The Wealth of Nations*, Adam Smith: Adam Smith, *An Inquiry into the Nature and Causes of the Wealth of Nations* (Edinburgh: Thomas Nelson, 1843), 95.

111 One writer noted that pigs could be fed: Markham, *Cheape and Good*, 106.

111 In 1621 a London maker of starch: Joan Thirsk, *Economic Policy and Projects* (Oxford: Clarendon Press, 1978), 91.

113 Alcohol production provided an even larger source: Peter Mathias, *The Brewing Industry in England* (Cambridge: Cambridge University Press, 1959), 42.

113 Daniel Defoe reported that Wiltshire and Gloucestershire produced: Daniel Defoe, *A Tour thro' the Whole Island of Great Britain* (London: Strahan, 1725), 48.

113 Dairymaids churned butter, and the whey flowed: Robert Malcolmson, *The English Pig* (London: Hambledon, 2001), 39.

113 The British navy required as many as 40,000 pigs: Daniel Baugh, *British Naval Administration in the Age of Walpole* (Princeton, NJ: Princeton University Press, 1965), 407–410.

113 Those legumes became hog feed: Adolphus Speed, *The Husbandman, Farmer, and Grasier's Compleat Instructor* (London: Henry Nelme, 1697), 86.

114 In many cases, these pigs reached slaughter weight: S. White, "From Globalized Pig Breeds to Capitalist Pigs," *Environmental History* 16 (2011): 103–104.

114 An English agricultural writer picked up on this: John Laurence, *A New System of Agriculture* (Dublin: J. Hyde, 1727), 100.

114 Analysis of mitochondrial DNA shows: E. Giuffra et al., "The Origin of the Domestic Pig," *Genetics* 154 (2000): 1788.

114 In Neolithic China swine had served as a key source: S. O. Kim, "Burials, Pigs, and Political Prestige in Neolithic China," *Current Anthropology* 35 (1994).

115 Even in the twentieth century, pork accounted: E. Anderson, *The Food of China* (New Haven, CT: Yale University Press, 1988), 177.

115 Overall, however, pork represented just a tiny part: John L. Buck, *Land Utilization in China* (New York: Paragon, 1968), 411.

116 In the words of Chairman Mao: F. Bray, "Agriculture," in *Science and Civilization in China*, ed. J. Needham (Cambridge: Cambridge University Press, 1984), 6:4.

116 It faced the problem of a growing population: Y. Yu, "Three Hundred Million Pigs," in *Feeding a Billion*, ed. S. H. Wittwer et al. (East Lansing: Michigan State University Press, 1987), 309–323.

116 This left no open land for pasturing: Earl B. Shaw, "Swine Industry of China," *Economic Geography* 14 (1938): 381–390.

116 When modernizers introduced American pig breeds: Sigrid Schmalzer, "Breeding a Better China: Pigs, Practices, and Place in a Chinese County, 1929–1937," *Geographical Review* 92 (2002): 17.

117 Some Chinese sows produced litters: H. Epstein, *Domestic Animals of China* (Farnham Royal, UK: Commonwealth Agricultural Bureau, 1969), 74.

CHAPTER 9

119 The ships also carried a menagerie: Charles Mann, *1493: Uncovering the New World Columbus Created* (New York: Knopf, 2011), 3–8.

120 These voyages started what has become known: Alfred Crosby, *The Columbian Exchange* (Westport, CT: Greenwood Press, 1972).

121 Just two years after Columbus's second expedition: R. A. Donkin, "The Peccary," *Transactions of the American Philosophical Society* 75 (1985): 41.

121 In a few more years the number of hogs running wild: Alfred Crosby, *Ecological Imperialism*, 2nd ed. (Cambridge: Cambridge University Press, 2004), 175.

121 "Do not kill them": Crosby, *Columbian Exchange*, 78.

121 **When the 150 people aboard made it to shore:** Virginia DeJohn Anderson, "Somer Islands' 'Hogge Money,'" *Environmental History* 9 (2004): 128–131.
121 **Columbus wrote that the trees and plants:** Alfred Crosby, "Metamorphosis of the Americas," in *Seeds of Change*, ed. Herman Viola and Carolyn Margolis (Washington, DC: Smithsonian Institution Press, 1991), 76.
122 **One Spaniard risked blasphemy by claiming:** B. J. Zadik, "The Iberian Pig in Spain and the Americas at the Time of Columbus" (master's thesis, University of California, Berkeley, 2005), 24.
122 **The Jamaican mountains soon held:** Donkin, "Peccary," 44.
122 **in 1514 the governor of Cuba told King Ferdinand:** Deb Bennett, "Ranching in the New World," in Viola and Margolis, *Seeds of Change*, 101.
122 **Peccaries, the American cousins to Eurasian pigs:** Lyle K. Sowls, *Javelinas and Other Peccaries* (College Station: Texas A&M University Press, 1997), 143–158.
122 **American societies had developed without the livestock:** Jared Diamond, *Guns, Germs, and Steel* (New York: W. W. Norton, 1998), 46–47.
123 **Spanish soldiers, brutal as they were:** Mann, *1493*, 97–101; F. Guerra, "The Earliest American Epidemic," *Social Science History* 12 (1988): 305–325; A. F. Ramenofsky and P. Galloway, "Disease and the Soto Entrada," in *Hernando de Soto Expedition*, ed. P. Galloway (Lincoln: University of Nebraska Press, 1997), 259–279.
124 **A Spanish historian has argued:** Crosby, *Columbian Exchange*, 77.
125 **His enemies called him a swineherd:** D. E. Vassberg, "Concerning Pigs, the Pizarros, and the Agro-Pastoral Background of the Conquerors of Peru," *Latin American Research Review* 13 (1978): 47–61.
125 **On sandy soils closer to the coast:** Angelos Hadjikoumis, "Traditional Pig Herding Practices in Southwest Iberia," *Journal of Anthropological Archaeology* 31 (2012): 353–364; T. Plieninger, "Constructed and Degraded? Origin and Development of the Spanish Dehesa Landscape," *Erde Berlin* 138 (2007): 25–46; D. W. Gade, "Parsons on Pigs and Acorns," *Geographical Review* 100 (2010): 598–606.
125 **Given the ample supply of mast for hogs:** James T. Parsons, "The Acorn-Hog Economy of the Oak Woodlands of Southwestern Spain," *Geographical Review* 52 (1962): 234.
126 **In 1554 one community in Extremadura reported:** Parsons, "Acorn-Hog," 215.
126 **Even that astonishing number wasn't enough:** Zadik, "Iberian Pig," 44–48.
127 **Only in dire circumstances would the leader:** Lawrence A. Clayton et al., *The De Soto Chronicles* (Tuscaloosa: University of Alabama Press, 1993), 81.
127 **When De Soto died of illness:** Clayton, *De Soto Chronicles*, 138–139.
128 **In most places pigs became village scavengers:** Lauren Derby, "Bringing the Animals Back In: Writing Quadrupeds into the Environmental History of Latin America and the Caribbean," *History Compass* 9 (2011): 605.

CHAPTER 10

131 **Spain derived its power from "Indian gold":** Anthony Pagden, *Lords of All the World* (New Haven, CT: Yale University Press, 1995), 67.
131 **The English began to describe the New World's gold:** Pagden, *Lords*, 68.
132 **The English saw themselves as fulfilling God's decree:** Genesis 1:26, KJV.

132 Britain built an empire to rival Spain's: Peter Coclanis, "Food Chains," *Agricultural History* 72 (2010): 667.

132 And Indians ultimately fared little better: The argument in this chapter derives largely from Virginia DeJohn Anderson, *Creatures of Empire* (New York: Oxford University Press, 2004).

133 Some sixty years before Raleigh: William Cronon, *Changes in the Land* (New York: Hill and Wang, 1983), 25.

133 Europeans marveled at the productivity: Charles Mann, *1491: New Revelations of the Americas Before Columbus* (New York: Knopf, 2005), 264–265; Emily Russell, *People and the Land Through Time* (New Haven, CT: Yale University Press, 1997), 26.

133 Columbus had been the first European to write: Betty Harper Fussell, *The Story of Corn* (New York: Knopf, 1992), 17.

133 William Wood, in Massachusetts, praised the Indian women: Anderson, *Creatures of Empire*, 81.

133 The Indians, wrote John Winthrop: Cronon, *Changes in the Land*, 130.

134 Robert Gray wrote that in Virginia: Anderson, *Creatures of Empire*, 79.

134 To justify seizing native land: Pagden, *Lords*, 76–79.

134 According to Roger Williams: Anderson, *Creatures of Empire*, 211.

134 In 1656, Virginia's legislators offered: Anderson, *Creatures of Empire*, 107.

135 As historian Virginia DeJohn Anderson has phrased it: Anderson, *Creatures of Empire*, 108.

136 Sheep, because of what one colonist called: Anderson, *Creatures of Empire*, 110.

137 "The real American hog," one observer said: L. C. Gray, *History of Agriculture in the Southern United States to 1860* (Gloucester, MA: Peter Smith, 1958), 206.

137 That was the toughness needed: Roger Williams, *A Key into the Language of America* (Bedford, MA: Applewood Books, 1997), 114.

139 Livestock had the legal right to all land: David Grettler, "Environmental Change and Conflict over Hogs in Early Nineteenth-Century Delaware," *Journal of the Early Republic* 19 (1999): 197–220.

139 America's farmers were "the most negligent": Anderson, *Creatures of Empire*, 244.

139 A more acute observer explained: Steven Stoll, *Larding the Lean Earth* (New York: Hill and Wang, 2002), 127.

139 One man in Virginia reported: Gray, *History of Agriculture*, 20.

139 A planter in Georgia explained: John Mitchell and Arthur Young, *American Husbandry* (London: J. Bew, 1775), 347.

139 Virginian Robert Beverley noted: Robert Beverley, *History and Present State of Virginia*, ed. Susan Scott Parrish (Chapel Hill: University of North Carolina Press, 2013), 251.

139 In 1660, Samuel Maverick reported: Cronon, *Changes in the Land*, 139.

139 As a Barbados planter explained: Anderson, *Creatures of Empire*, 152; John Otto, *The Southern Frontiers, 1607–1860* (New York: Greenwood Press, 1989), 33.

139 Pork and beef became New England's: Anderson, *Creatures of Empire*, 152.

140 On the eve of the American Revolution: John McCusker, *The Economy of British America, 1607–1789* (Chapel Hill: University of North Carolina Press, 1985), 268.

141 "Tis true indeed, none of my deer are marked": Anderson, *Creatures of Empire*, 216–217.

141 **Often pigs were simply pushed further away:** Percy Bidwell and John Falconer, *History of Agriculture in the Northern United States, 1620–1860* (New York: P. Smith, 1941), 22.

142 **They devoured tuckahoe, a starchy root:** Gordon Whitney, *From Coastal Wilderness to Fruited Plain* (Cambridge: Cambridge University Press, 1996), 165.

142 **Roger Williams observed that pigs lingered:** Williams, *Key*, 114.

142 **A more likely reason Indians disliked swine:** Cotton Mather, *Magnalia Christi Americana* (Hartford, CT: Silas Andeus, 1853), 1:560.

143 **"But these English having gotten our land":** Anderson, *Creatures of Empire*, 207.

143 **Mattagund, an Indian leader in Maryland:** Anderson, *Creatures of Empire*, 221.

CHAPTER 11

145 **But European settlers had arrived with livestock:** William Bowen, *The Willamette Valley* (Seattle: University of Washington Press, 1978), 87.

146 **The American settlers did so using:** Terry Jordan-Bychkov, *The American Backwoods Frontier* (Baltimore: Johns Hopkins University Press, 1989), 123; also see Steven Stoll, *Larding the Lean Earth* (New York: Hill and Wang, 2002), 104.

146 **In 1823 New England traveler Timothy Dwight:** Timothy Dwight, *Travels in New-England and New-York* (London: W. Barnes, 1823), 2:439.

146 **One German observer noted in the 1780s:** Jordan-Bychkov, *American Backwoods*, 4.

146 **"Of all the domestic animals":** Reuben Gold Thwaites et al., *Early Western Travels, 1748–1846* (Cleveland, OH: Clark, 1905), 3:246.

146 **A traveler in Ohio in 1817 reported:** Silas Chesebrough, "Journal of a Journey to the Westward," *American Historical Review* 37 (1931): 82–83.

147 **Fordham encouraged Englishmen to seek their fortunes:** Elias Pym Fordham, *Personal Narrative of Travels in Virginia, Maryland . . .* (Cleveland, OH: Arthur H. Clark Co., 1906), 120, 236.

147 **The western poet Charles Badger Clark captured:** Charles Badger Clark, *Sun and Saddle Leather* (Boston: R. G. Badger, 1920), 75.

147 **In early Ohio, one man observed:** Robert Leslie Jones, *History of Agriculture in Ohio to 1880* (Kent, OH: Kent State University Press, 1983).

148 **"We put shelled corn in the pen":** Oliver Johnson and Howard Johnson, *A Home in the Woods* (Bloomington: Indiana University Press, 1978), 109.

149 **Abraham Lincoln described himself:** William Barton, *The Soul of Abraham Lincoln* (New York: George H. Doran, 1920), 53.

149 **Woods pigs were called razorbacks:** Jones, *History of Agriculture in Ohio*, 121; Rudolf Clemen, *The American Livestock and Meat Industry* (New York: Ronald Press, 1923), 53; John Mack Faragher, *Sugar Creek* (New Haven, CT: Yale University Press, 1986), 65; Allan G. Bogue, *From Prairie to Corn Belt* (Chicago: University of Chicago Press, 1963), 105; Mart Stewart, *"What Nature Suffers to Groe": Life, Labor, and Landscape on the Georgia Coast, 1680–1920* (Athens: University of Georgia Press, 1996), 213; Robert Porter, William Jones, and Henry Gannett, *The West from the Census of 1880* (Chicago: Rand McNally, 1882), 309.

149 **"Drops of fat dripped off it":** Laura Ingalls Wilder, *Little House in the Big Woods* (New York: HarperTrophy, 1971), 15–16.

149 **"In all my previous life":** Sam Bowers Hilliard, *Hog Meat and Hoecake* (Carbondale: Southern Illinois University Press, 1972), 39.

149 **Frederick Law Olmsted, a journalist:** Frederick Law Olmsted, *A Journey Through Texas* (New York: Dix, Edwards, 1857), 15.

150 **Then he added a lament familiar:** George William Featherstonhaugh, *Excursion Through the Slave States* (London: J. Murray, 1844), 2:109.

150 **"The ordinary mode of living is abundant":** Frances Trollope, *Domestic Manners of the Americans* (London: Whittaker, Treacher, 1832), 238.

150 **Sites in the Ozarks dating to a few decades:** Samuel Smith, *Historical Background and Archaeological Testing of the Davy Crockett Birthplace State Historical Area* (Nashville: Tennessee Department of Conservation, 1980); C. R. Price and J. E. Price, "Investigation of Settlement and Subsistence Systems in the Ozark Border Region of Southeast Missouri During the First Half of the 19th Century," *Ethnohistory* 28 (1981): 237–258.

150 **They raised swine during the early years:** Brian Hesse, "Pig Lovers and Pig Haters," *Journal of Ethnobiology* 10 (1990): 218.

151 **In later centuries, once the herds:** Pam Crabtree, "Sheep, Horses, Swine, and Kine," *Journal of Field Archaeology* 16 (1989): 205–213.

151 **"emancipated themselves from":** Theodore Blegen, *Norwegian Migration to America* (Northfield, MN: Norwegian-American Historical Association, 1931), 195.

CHAPTER 12

153 **"Here, in Ohio, they are intelligent":** Reuben Gold Thwaites et al., *Early Western Travels, 1748–1846* (Cleveland, OH: Clark, 1905), 19:33.

154 **It became, and remains, the agricultural heartland:** John Hudson, *Making the Corn Belt* (Bloomington: Indiana University Press, 1994), 58–59.

154 **One scholar estimates that if Americans:** Terry Jordan-Bychkov, *The American Backwoods Frontier* (Baltimore: Johns Hopkins University Press, 1989), 115.

156 **The land, the Renicks wrote:** Hudson, *Making the Corn Belt*, 60.

156 **One of the Renicks later described their system:** Hudson, *Making the Corn Belt*, 68.

157 **After the cows had eaten:** Hudson, *Making the Corn Belt*, 71; for medieval use of this feeding technique, see Irven Resnick, *Marks of Distinction* (Washington, DC: Catholic University of America Press, 2012), 170.

157 **As historian Allan Bogue has explained:** Allan G. Bogue, *From Prairie to Corn Belt* (Chicago: University of Chicago Press, 1963), 103.

158 **A book on the early years of the Corn Belt observes:** Paul Henlein, *Cattle Kingdom in the Ohio Valley, 1783–1860* (Lexington: University Press of Kentucky, 1959), 73.

158 **"Hogs don't always carry the prestige":** Hudson, *Making the Corn Belt*, 74.

158 **The hog earned the nickname:** Hudson, *Making the Corn Belt*, 74.

158 **"What is a hog":** James Parton, "Chicago," *Atlantic Monthly* 19 (1867): 331; H. C. Hill, "The Development of Chicago as a Center of the Meat Packing Industry," *Mississippi Valley Historical Review* 10 (1923): 260.

159 **In 1790 an English agriculture writer:** William Marshall, *The Rural Economy of the Midland Counties* (London: G. Nicol, 1790), 453.

159 **A swine expert noted that the new types:** S. White, "From Globalized Pig Breeds to Capitalist Pigs," *Environmental History* 16 (2011): 108.

160 **Only the second half of the breed's name:** Josiah Morrow, *The History of Warren County, Ohio* (Chicago: W. H. Beers, 1882), 323–324.

160 **In 1840 there were more than 26 million:** *United States Census of Agriculture: 1950* (Washington, DC: Bureau of the Census, 1951), 2:362; *Statistics of the United Kingdom of Great Britain and Ireland* (London: Thom Alexander, 1868), 42.

160 **In improved Corn Belt hogs:** Alan Olmstead, *Creating Abundance* (New York: Cambridge University Press, 2008), 312–313.

160 **Whereas woods hogs took two or three years:** Margaret Walsh, *The Rise of the Midwestern Meat Packing Industry* (Lexington: University Press of Kentucky, 1982), 23.

160 **"Nowhere in the world can such marvelous herds":** H. J. Carman, "English Views of Middle Western Agriculture, 1850–1870," *Agricultural History* 8 (1934): 17–18.

161 **An agricultural newspaper explained:** Rudolf Clemen, *The American Livestock and Meat Industry* (New York: Ronald Press, 1923), 58n18.

161 **The Poland China dominated the early Corn Belt:** Hudson, *Making the Corn Belt*, 84.

161 **"That was the prettiest drive of anything":** William Lynwood Montell, *Don't Go Up Kettle Creek* (Knoxville: University of Tennessee Press, 1983), 45–46; E. Coulter, *Auraria* (Athens: University of Georgia Press, 2009), 21.

161 **The best estimates suggest that in antebellum America:** Sam Bowers Hilliard, *Hog Meat and Hoecake* (Carbondale: Southern Illinois University Press, 1972), 195.

161 **In 1847 one tollgate in North Carolina recorded:** *Highland Messenger* (Asheville, NC), January 14, 1842.

162 **A few farmers from Lexington, Kentucky:** Elizabeth Parr, "Kentucky's Overland Trade with the Ante-Bellum South," *Filson Club Quarterly* 2 (1928): 72.

163 **The drivers shouted, "Soo-eey":** Edmund Burnett, "Hog Raising and Hog Driving in the Region of the French Broad River," *Agricultural History* 20 (1946): 90.

163 **The secret, one drover said:** Montell, *Don't Go*, 42.

163 **The young Abraham Lincoln:** William Barton, *The Soul of Abraham Lincoln* (New York: George H. Doran, 1920), 46.

164 **We don't know many details:** Michael Ross MacKinnon, "Animal Production and Consumption in Roman Italy" (PhD diss., University of Alberta, 1999), 130–131.

164 **One traveler described watching a drove:** Thomas Searight, *The Old Pike* (Uniontown, PA: T. Searight, 1894), 142–143.

164 **The largest cattle drives, from Texas to Kansas:** Richard White, "Animals and Enterprise," in *The Oxford History of the American West*, ed. Clyde Milner et al. (New York: Oxford University Press, 1994), 260.

164 **From Kentucky alone, as many as 100,000 hogs:** Frederick Jackson Turner, *Rise of the New West, 1819–1829* (New York: Harper, 1906), 101.

164 **In 1855 more than 83,000 hogs:** *Asheville News*, February 1, 1855.

164 **The route through the Cumberland Gap:** Dwight Billings, *The Road to Poverty* (New York: Cambridge University Press, 2000), 47.

CHAPTER 13

167 **Americans, she thought, were overconfident and undereducated:** Frances Trollope, *Domestic Manners of the Americans* (London: Whittaker, Treacher, 1832), 12.

167 **"I am sure I should have liked Cincinnati":** Trollope, *Domestic Manners*, 85.

168 **As she was on a stroll one day:** Trollope, *Domestic Manners*, 85.

168 **"'Tis to be a slaughter-house for hogs":** Trollope, *Domestic Manners*, 98.

169 **Dozens of midsized packers were scattered:** Margaret Walsh, *The Rise of the Midwestern Meat Packing Industry* (Lexington: University Press of Kentucky, 1982), 41.

169 **Just as importantly, the rivers allowed salt:** Isaac Lippincott, "The Early Salt Trade of the Ohio Valley," *Journal of Political Economy* 20 (1912): 1034–1035.

170 **By the 1870s, it had reached 6 million:** Margaret Walsh, "Pork Packing as a Leading Edge of Midwestern Industry, 1835–1875," *Agricultural History* 51 (1977): 704.

170 **"hog butcher for the world":** Carl Sandburg, "Chicago," in *Complete Poems of Carl Sandburg* (New York: Houghton Mifflin Harcourt, 2003), 3.

170 **slaughtering as many as 4 million hogs:** Walsh, *Rise*, 8.

170 **With the coming of the Civil War:** William Cronon, *Nature's Metropolis* (New York: W. W. Norton, 1991), 229–230.

170 **In 1838, after a visit to Cincinnati:** Harriet Martineau, *Retrospect of Western Travel* (London: Saunders and Otley, 1838), 233.

170 **a reference to a famous passage in *The Wealth of Nations*:** Adam Smith, *An Inquiry into the Nature and Causes of the Wealth of Nations* (Edinburgh: Thomas Nelson, 1843), 3–4.

170 **Twenty years later Frederick Law Olmsted described:** Frederick Law Olmsted, *A Journey Through Texas* (New York: Dix, Edwards, 1857), 9.

172 **Each worker had just twelve seconds:** James Parton, "Cincinnati," *Atlantic Monthly* 20 (1867): 240–243; Charles Cist, "The Hog and Its Products," in *Report of the Commissioner of Agriculture for the Year 1866*, ed. J. W. Stokes (Washington, DC: Government Printing Office, 1867), 392–396; Charles Cist, *Cincinnati Miscellany* (Cincinnati, OH: C. Clark, 1845).

172 **There were two ways to make it more efficient:** S. Giedion, *Mechanization Takes Command* (New York: Oxford University Press, 1948), 93.

173 **The genius of the packers' disassembly line:** Henry Ford, *My Life and Work* (Garden City, NY: Doubleday, 1922), 80–81; David E. Nye, "What Was the Assembly Line?," *Tidsskrift for Historie* 1 (2010): 59–81.

173 **"Great as this wonderful city is in everything":** Cronon, *Nature's Metropolis*, 207.

173 **In Upton Sinclair's *The Jungle*, a slaughterhouse employee:** Upton Sinclair, *The Jungle* (New York: Penguin: 2001), 38.

175 **In 1837 corn-fed hogs sold for $5:** Rudolf Clemen, *The American Livestock and Meat Industry* (New York: Ronald Press, 1923), 54.

175 **A visitor to the Illinois prairie in 1837:** S. A. Mitchell, *Illinois in 1837* (Philadelphia: S. A. Mitchell, 1837), 42.

176 **Organs, as well as meat from ribs and necks:** Cist, "Hog," 386.

177 **That meant big packers could pay higher prices for pigs:** Cist, "Hog," 385–386.

177 **Widows typically received 120 pounds:** Richard Cummings, *The American and His Food* (Chicago: University of Chicago Press, 1941), 15.

177 **In the antebellum South, a typical ration:** Kenneth Kiple, *Another Dimension to the Black Diaspora* (New York: Cambridge University Press, 1981), 81–82.

177 **Laborers in the North ate 170 pounds:** W. J. Warren, *Tied to the Great Packing Machine* (Iowa City: University of Iowa Press, 2007), 221.

177 **"There are a great many ill conveniences here":** Arthur Schlesinger, *Paths to the Present* (New York: Houghton Mifflin, 1964), 244.

177 **The United States in 1900 saw itself as a nation:** Warren, *Tied*, 223.

178 **and even in 1900 only two:** Roderick Floud, *The Changing Body* (New York: Cambridge University Press, 2011), 309.

178 **In his 1845 novel *The Chainbearer:*** James Fenimore Cooper, *The Chainbearer* (New York: D. Appleton, 1833), 102.

178 **In Eliza Leslie's *Directions for Cookery:*** Eliza Leslie, *Directions for Cookery* (Philadelphia: Carey & Hart, 1844), 120.

178 **One man, recalling his midwestern childhood:** Robert Leslie Jones, *History of Agriculture in Ohio to 1880* (Kent, OH: Kent State University Press, 1983).

178 **As a physician wrote in the magazine *Godey's Lady's Book:*** Joseph Stainback Wilson, "Quantity of Gastric Juice," *Godey's Lady's Book* 60 (1860): 178.

179 **The upper class bought refrigerated beef:** Richard Perren, *Taste, Trade and Technology* (Burlington, VT: Ashgate, 2006), 1–8.

179 **As shipping technology improved and global trade expanded:** Jack Goody, *Cooking, Cuisine, and Class* (New York: Cambridge University Press, 1982), 154.

179 **Workers who once had spent 50 to 75 percent:** Stephen Broadberry and Kevin 179. O'Rourke, *The Cambridge Economic History of Modern Europe* (New York: Cambridge University Press, 2010), 1:148; Wilhelm Abel, *Agricultural Fluctuations in Europe* (London: Methuen and Company, 1978), 142.

180 **The average height of adults:** Gretel Pelto and Pertti Pelto, "Diet and Delocalization," *Journal of Interdisciplinary History* 14 (1983): 514–515.

CHAPTER 14

181 **As they rounded up stray pigs:** Catherine McNeur, "'The Swinish Multitude': Controversies over Hogs in Antebellum New York City," *Journal of Urban History* 37 (2011): 639.

181 **Someone threw a brick:** McNeur, "Swinish Multitude," 639.

181 **The city's streets functioned:** Hendrik Hartog, "Pigs and Positivism," *Wisconsin Law Review* (July/August 1985): 904.

182 **The odds of that happening were good:** McNeur, "Swinish Multitude," 640.

182 **The pigs devoured "all kinds of refuse":** Ole Munch Ræder, *America in the Forties* (Minneapolis: University of Minnesota Press, 1929), 78.

182 **Colonial New England towns appointed "hog reeves":** William Cronon, *Changes in the Land* (New York: Hill and Wang, 1983), 136.

183 **These were not chubby Corn Belt pigs:** Charles Dickens, *American Notes* (London: Chapman and Hall, 1868), 38.

183 **Loose pigs in the streets:** C. H. Wilson, *The Wanderer in America* (Thirsk, UK: H. Masterman, 1822), 18.

183 **In *The Condition of the Working-Class in England*:** Friedrich Engels, *The Condition of the Working-Class in England in 1844* (London: Allen and Unwin, 1892), 52–53, 49.

183 **In the Potteries, a large slum:** Peter Stallybrass, *The Politics and Poetics of Transgression* (Ithaca, NY: Cornell University Press, 1986), 147.

183 **In *Reflections on the Revolution in France:*** Edmund Burke, *Reflections on the Revolution in France* (London: J. Dodsley, 1790), 117.

184 **In the 1860s and 1870s, public health measures:** Robert Malcolmson, *The English Pig* (London: Hambledon, 2001), 43.

184 **New York's professional police force:** McNeur, "Swinish Multitude," 648.

185 **Southerners "delight in their present low":** Forrest McDonald and Grady McWhiney, "The South from Self-Sufficiency to Peonage: An Interpretation," *American Historical Review* 85 (1980): 1095.

185 **If this were true:** R. Ben Brown, "The Southern Range: A Study of Nineteenth Century Law and Society" (PhD diss., University of Michigan, 1993), 161.

186 **That left millions of acres available:** Brown, "Southern Range," 3; Michael Williams, *Americans and Their Forests* (New York: Cambridge University Press, 1989), 118–120.

186 **"You can keep as many pigs as you wish":** Joseph Eder, "A Bavarian's Journey to New Orleans and Nacogdoches in 1853–1854," *Louisiana Historical Quarterly* 23 (1940): 497.

186 **A prosperous farmer in the Blue Ridge Mountains:** M. R. Walpole, "The Closing of the Open Range in Watauga County, NC," *Appalachian Journal* 16 (1989): 326.

186 **Barbecue—both the word and the technique were borrowed:** John Shelton Reed, Dale Volberg Reed, and William McKinney, *Holy Smoke* (Chapel Hill: University of North Carolina Press, 2008), 12–14.

186 **As one English visitor to America explained:** Reuben Gold Thwaites et al., *Early Western Travels, 1748–1846* (Cleveland, OH: Clark, 1905), 11:106.

187 **Laborers, he wrote, if "furnished with free food":** Brown, "Southern Range," 190.

187 **Legislatures in every state closed the range:** Brown, "Southern Range," 280.

187 **A newspaper claimed that this change:** Brown, "Southern Range," 217.

187 **This was indeed the effect:** E. P. Thompson, *The Making of the English Working Class* (New York: Pantheon, 1964), 217.

187 **Historians who examined six counties in Alabama and Mississippi:** McDonald and McWhiney, "South from Self-Sufficiency," 1114.

188 **In the 1880s, one writer described:** Brown, "Southern Range," 222.

188 **In nineteenth-century England:** Peter Bowden, "Agricultural Prices, Farm Profits, and Rents," in *The Agrarian History of England and Wales*, ed. Joan Thirsk and H. P. R. Finberg (Cambridge: Cambridge University Press, 1967), 4:416.

188 **"Life without a pig was almost unthinkable":** Walter Rose, *Good Neighbours* (Cambridge: Cambridge University Press, 2010), 58.

188 **The pig was "one of the best friends of the poor":** Malcolmson, *English Pig*, 45.

188 **"The pig was an important member of the family":** Flora Thompson, *Lark Rise to Candleford* (Boston: David R. Godine, 2009), 10.

189 **"Not much profit there":** Ralph Whitlock, *The Land First* (London: Museum Press, 1954), 78.

189 **"Watching her now as she tucked into a sort of hash":** P. G. Wodehouse, *Heavy Weather*, in *Life at Blandings* (New York: Penguin, 1981), 415.

189 **E. B. White, who lived:** E. B. White, "Death of a Pig," in *The American Idea*, ed. Robert Vare (New York: Doubleday, 2007), 286–294.

190 According to an English observer, "A man": Reginald Ernest Moreau, *The Departed Village* (London: Oxford University Press, 1968), 114.

190 A study of Oxfordshire suggested that raising pigs: Malcolmson, *English Pig*, 57.

190 "Pig clubs," a sort of mutual insurance program: Malcolmson, *English Pig*, 59.

190 In *Middlemarch*, George Eliot defines a happy village: George Eliot, *Middlemarch* (New York: Harper & Brothers, 18732), 277.

191 "The killing of the pig": Rose, *Good Neighbours*, 65.

191 Pa removes the bladder: Laura Ingalls Wilder, *Little House in the Big Woods* (New York: Harper Trophy, 1971), 14–15.

191 One girl recalled that, during the pig killing: Thompson, *Lark Rise*, 12, 271.

191 Thomas Hardy devotes an entire chapter of *Jude the Obscure:* Thomas Hardy, *Jude the Obscure* (New York: Harper & Brothers, 1895), 1:71–72.

CHAPTER 15

195 "The ear was assailed by a most terrifying shriek": Upton Sinclair, *The Jungle* (New York: Penguin: 2001), 39–40.

196 A newspaper described Sinclair's concern for pig suffering: W. J. Warren, *Tied to the Great Packing Machine* (Iowa City: University of Iowa Press, 2007), 127.

196 "I aimed at the public's heart": James Harvey Young, *Pure Food* (Princeton, NJ: Princeton University Press, 1989), 229.

196 There was never the least attention paid: Sinclair, *Jungle*, 136–137.

198 The government victory proved only nominal: A. M. Azzam and Dale G. Anderson, *Assessing Competition in Meatpacking* (Washington, DC: US Department of Agriculture, 1996), 15–16; Alfred Chandler, *Strategy and Structure* (Cambridge, MA: MIT Press, 1969), 25–26.

198 Meatpackers had been using borax: Roger Horowitz, *Putting Meat on the American Table* (Baltimore: Johns Hopkins University Press, 2006), 59.

199 The federal government took over inspection: Doris Kearns Goodwin, *The Bully Pulpit* (New York: Simon & Schuster, 2013), 465–466.

199 In the decade after *The Jungle*'s publication: Warren, *Tied*, 145–146, 221.

199 Meatpackers, though, blamed *The Jungle:* Warren, *Tied*, 146.

199 According to Edward Hitchcock: Edward Hitchcock, *Dyspepsy Forestalled* (Amherst, MA: J. S. & C. Adams, 1831), 185.

199 "Fat bacon and pork are peculiarly appropriate for negroes": Joseph Stainback Wilson, "Quantity of Gastric Juice," *Godey's Lady's Book* 60 (1860): 178.

199 This was thanks in no small part to Sylvester Graham: Stephen Nissenbaum, *Sex, Diet, and Debility in Jacksonian America* (Westport, CT: Greenwood Press, 1980), 44.

200 America had inherited from England a hierarchy: Joan Thirsk, *Food in Early Modern England* (London: Continuum, 2006), 249.

200 One cookbook writer dismissed barreled pork: Horowitz, *Putting Meat*, 45.

200 another described pork as "dangerously unwholesome": Keith Stavely, *America's Founding Food* (Chapel Hill: University of North Carolina Press, 2004), 193–194.

200 An 1893 guide to household management claimed: Marion Harland, *Common Sense in the Household* (New York: Charles Scribner, 1893), 116.

200 For southern whites, the same was true: Horowitz, *Putting Meat*, 12.

201 American hams—especially Virginia's Smithfield variety: Joseph Earl Dabney, *Smokehouse Ham, Spoon Bread and Scuppernong Wine* (Nashville, TN: Cumberland House, 1998), 189.

201 Later they invented "vein-pumping": Horowitz, *Putting Meat*, 60–62.

202 The method could not deliver the intense flavor: Horowitz, *Putting Meat*, 58.

202 By 1960 bacon had shed its reputation: Horowitz, *Putting Meat*, 62–69.

202 leading to headlines such as: "Missouri Town Reports 47 Cases of Trichinosis," *New York Times*, January 28, 1969.

203 When in-sink garbage grinders such as the DisposAll: "Garbage Grinder Becomes an Issue," *New York Times*, May 5, 1966.

203 In a period of just over two years: Thomas Moore, "Prevailing Methods of Garbage Collection and Disposal in American Cities," *The American City* 22 (1920): 602–608; Martin Melosi, *The Sanitary City* (Pittsburgh, PA: University of Pittsburgh Press, 2008), 116, 127, 163, 206.

203 The farms survived until 1960: Orville Schell, *Modern Meat* (New York: Random House, 1984), 71–85.

204 "Human trichinosis is based almost entirely on porcine trichinosis": Sylvester Gould, *Trichinosis in Man and Animals* (Springfield, IL: Thomas, 1970), 506.

204 "Although garbage-fed hogs are daily sold as food": Moore, "Prevailing Methods."

204 Meat quality suffered: R. Lawrie, *Lawrie's Meat Science* (Boca Raton, FL: Woodhead Publishing, 2006), 99.

204 As one scientific study noted: Gould, *Trichinosis* 394.

204 A 1942 study from the USDA noted: *Family Food Consumption in the United States, Spring 1942* (Washington, DC: US Department of Agriculture, 1944), 15.

205 A 1955 study of urban consumers found: Faith Clark et al., *Food Consumption of Urban Families in the United States* (Washington, DC: US Department of Agriculture, 1955), Table 75.

205 Urbanites devoted half of their meat consumption to beef: *Family Food Consumption*, 14–15.

205 That year, for the first time, Americans ate more beef: J. L. Anderson, "Lard to Lean: Making the Meat-Type Hog in Post–World War II America," in *Food Chains*, ed. Warren Belasco and Roger Horowitz (Philadelphia: University of Pennsylvania Press, 2009), 30.

205 By the 1970s, pork consumption had fallen: Warren, *Tied*, 222; Christopher Davis and Biing-Hwan Lin, *Factors Affecting U.S. Pork Consumption* (Washington, DC: US Department of Agriculture, 2005), 4.

205 The president of the National Pork Producers Council traveled the country: Don Muhm, *Iowa Pork and People* (Clive: Iowa Pork Foundation, 1995), 60.

205 They created a mascot, Lady Loinette: Jenny Barker Devine, "'Hop to the Top with the Iowa Chop': The Iowa Porkettes and Cultivating Agrarian Feminisms in the Midwest, 1964–1992," *Agricultural History* 83 (2009): 480.

206 One queen asked, "Who first but Iowa": Muhm, *Iowa Pork*, 100.

206 The Porkettes held contests for baking with lard: Muhm, *Iowa Pork*, 90.

206 The group's magazine, *Ladies Pork Journal*, included: Muhm, *Iowa Pork*, 79.

206 The first president of the Porkettes told a story: Devine, "Hop to the Top," 478.

206 Another Porkette was conducting a grocery store promotion: Muhm, *Iowa Pork*, 107.

CHAPTER 16

207 **When the lights came on:** Rolland Paul et al., *The Pork Story* (Des Moines, IA: NPPC, 1991), 183.

207 **Many thought it was a "dumb idea":** Paul et al., *Pork Story*, 183.

208 **In one survey more than a third of Americans agreed:** J. L. Anderson, "Lard to Lean: Making the Meat-Type Hog in Post–World War II America," in *Food Chains*, ed. Warren Belasco and Roger Horowitz (Philadelphia: University of Pennsylvania Press, 2009), 42; also see National Research Council, *Designing Foods* (Washington, DC: National Academy Press, 1988), 43.

208 **Eight out of ten Americans recognized the phrase:** "Humane Society Lawsuit Brings National Pork Board Response," *Western Farm Press*, September 27, 2012; Richard Horwitz, *Hog Ties* (Minneapolis: University of Minnesota Press, 2002), 39.

208 **In 2011 *Adweek* deemed the campaign:** Ed Norton, "'The Other White Meat' Finally Cedes Its Place," *Adweek*, March 4, 2011.

210 **In 1907 the Danes had created swine testing stations:** Earl B. Shaw, "Swine Industry of Denmark," *Economic Geography* 14 (1938): 23; Julian Wiseman, *The Pig*, 2nd ed. (London: Duckworth, 2000), 71–73.

210 **The Hampshire registry, for instance, specified:** Don Muhm, *Iowa Pork and People* (Clive: Iowa Pork Foundation, 1995), 196.

211 **By the 1970s, a pig of the same size:** Anderson, "Lard to Lean," 39.

211 **They roamed on pasture in the spring:** F. B. Morrison, *Feeds and Feeding* (Ithaca, NY: Morrison Publishing, 1956), 843–867.

211 **In 1938 the United States raised 62 million hogs:** Richard Perren, *Taste, Trade and Technology* (Burlington, VT: Ashgate, 2006), 110; Earl B. Shaw, "Swine Production in the Corn Belt of the United States," *Economic Geography* 12 (1936): 359.

211 **Since the 1920s scientists had understood:** Terry G. Summons, "Animal Feed Additives, 1940–1966," *Agricultural History* 42 (1968): 305–306.

212 **Pfizer claimed that pigs dosed with Terramycin:** J. L. Anderson, *Industrializing the Corn Belt* (DeKalb, IL: Northern Illinois University Press, 2009), 93.

212 **By the 1960s, livestock consumed:** William Boyd, "Making Meat: Science, Technology, and American Poultry Production," *Technology and Culture* 42 (2001): 648; also see M. Finlay, "Hogs, Antibiotics and the Industrial Environments of Postwar Agriculture," in *Industrializing Organisms*, ed. Philip Scranton and Susan Schrepfer (New York: Routledge, 2004), 239.

213 **With farmland so expensive, one farmer asked:** Orville Schell, *Modern Meat* (New York: Random House, 1984), 61.

213 **Antibiotics, the drug salesman said, help pigs:** Schell, *Modern Meat*, 13.

213 **In 1972 an agricultural magazine predicted:** Anderson, *Industrializing*, 103.

213 **Farmhands, though, were scarce in the 1950s:** Paul Robbins, "Labor Situations Facing the Producer," in *The Pork Industry*, ed. David Topel (Ames: Iowa State University Press, 1968), 211.

214 **Compared to its ancestor in the 1930s, a broiler chicken:** Boyd, "Making Meat," 638.

214   Thanks to low prices and a healthy image: W. J. Warren, *Tied to the Great Packing Machine* (Iowa City: University of Iowa Press, 2007), 223.

214   "The use of slotted floors has probably accelerated": Al Jensen, *Management and Housing for Confinement Swine Production* (Urbana: University of Illinois, 1972), 7.

215   For each pig weighing 150 to 250 pounds: Wilson G. Pond, "Modern Pork Production," *Scientific American* 248 (1983): 102.

215   In such close quarters, pigs kept each other warm: Abigail Woods, "Rethinking the History of Modern Agriculture: British Pig Production, c. 1910–65," *Twentieth Century British History* 23 (2011): 173.

215   Crowded together, they shuffled around more: Chris Mayda, "Pig Pens, Hog Houses, and Manure Pits: A Century of Change in Hog Production," *Material Culture* 36 (2004): 25.

215   Slatted floors made the farmer's life easier: Ronald Plain, James R. Foster, and Kenneth A. Foster, *Pork Production Systems with Business Analyses* (Raleigh: North Carolina Cooperative Extension Service, 1995).

215   The "comfort and convenience" of the farmer: Jensen, *Management*, 3.

215   In a celebratory cover story in *Scientific American*: Pond, "Modern Pork Production," 96.

216   By 2000, three-quarters of sows in the United States: Nigel Key and William D. McBride, *The Changing Economics of U.S. Hog Production* (Washington, DC: US Department of Agriculture, 2007), 12.

216   Breeding stock tended to be purebreds: Paul Brassley, "Cutting Across Nature? The History of Artificial Insemination in Pigs in the United Kingdom," *Studies in History and Philosophy of Biological and Biomedical Sciences* 38 (2007): 442–461.

217   "If a sow has a litter of twelve and rolls on three": Schell, *Modern Meat*, 63.

217   Today, it takes less than three pounds: John McGlone, "Swine," in *Animal Welfare in Animal Agriculture*, ed. Wilson G. Pond, Fuller W. Bazer, and Bernard E. Rollin (Boca Raton, FL: CRC Press, 2012), 150.

217   As one animal scientist explained: McGlone, "Swine," 150.

218   Scientists from Texas A&M University showed: Anderson, "Lard to Lean," 43–44.

218   After an initial boost spurred by the campaign: Jane L. Levere, "The Pork Industry's 'Other White Meat' Campaign Is Taken in New Directions," *New York Times*, March 4, 2005; Trish Hall, "And This Little Piggy Is Now on the Menu," *New York Times*, November 13, 1991.

218   Meanwhile, consumption of chicken: Warren, *Tied*, 222.

218   At an industry conference in the 1960s: Irvin Omtvedt, "Some Heritability Characteristics," in Topel, *Pork Industry*, 128.

218   In 2000, industry experts writing in *National Hog Farmer:* Tom J. Baas and Rodney Goodwin, "Genetic-Based Niche Marketing Programs," *National Hog Farmer*, August 1, 2000.

219   "Their personalities are completely different": Temple Grandin and Catherine Johnson, *Animals in Translation* (New York: Scribner, 2005), 101. Also see Pond, "Modern Pork Production," 100.

219   As a group of veterinarians explained: European Union, *The Welfare of Intensively Kept Pigs* (Brussels: European Commission, 1997), 4.8.2.

CHAPTER 17

221 **"If it is not feasible to do this in a confinement operation"**: Bernard Rollin, *Putting the Horse Before Descartes* (Philadelphia: Temple University Press, 2011), 207.

222 **The *New York Times* printed its first analysis of modern pig farming**: William Serrin, "Hog Production Swept by Agricultural Revolution," *New York Times*, August 11, 1980.

222 **In 1995 the *Raleigh News and Observer* earned a Pulitzer Prize**: Pat Stith, Joby Warrick, and Melanie Sill, "Boss Hog," *Raleigh News and Observer*, February 19, 1995.

222 **As one industry insider explained, "For modern agriculture"**: Peter Cheeke, *Contemporary Issues in Animal Agriculture* (Danville, IL: Interstate, 1999), 248.

223 **By the late 1960s, however, a livestock expert had already noted**: Herrell DeGraff, "Introduction," in *The Pork Industry*, ed. David Topel (Ames: Iowa State University Press, 1968), xii.

223 **Secretary of Agriculture Ezra Taft Benson infamously offered this advice**: Wendell Berry, *The Way of Ignorance* (New York: Counterpoint, 2006), 117.

223 **In 1950, the average hog farm had 19 animals**: National Animal Health Monitoring System, *Swine 2006, Part IV* (Fort Collins, CO: US Department of Agriculture, 2008), 5.

223 **By 2004, 80 percent of hogs lived**: Nigel Key and William D. McBride, *The Changing Economics of U.S. Hog Production* (Washington, DC: US Department of Agriculture, 2007), 5.

223 **In 2010, the top four hog producers had captured two-thirds of the market**: Timothy A. Wise and Sarah E. Trist, "Buyer Power in U.S. Hog Markets," *Global Development and Environment Institute Working Papers* 10 (2010): 4, 6.

224 **The company owns the slaughtering plants**: Pew Commission on Industrial Farm Animal Production, *Putting Meat on the Table* (Philadelphia: Pew Charitable Trusts, 2008), 42.

224 **"Vertical integration gives you high-quality"**: David Barboza, "Goliath of the Hog World," *New York Times*, April 7, 2000.

224 **The number of hogs raised there doubled**: Key and McBride, *Changing Economics*, 9.

224 **Wendell Murphy, founder of a large hog producer**: Pat Stith and Joby Warrick, "Murphy's Law," *Raleigh News and Observer*, February 22, 1995; Michael Thompson, "This Little Piggy Went to Market: The Commercialization of Hog Production in Eastern North Carolina from William Shay to Wendell Murphy," *Agricultural History* 74 (2000): 569–584.

224 **By 2006, 95 percent of hogs in the United States**: Wise and Trist, "Buyer Power," 4, 6.

224 **1,300 hogs per hour in some cases**: Ted Genoways, *The Chain* (New York: Harper, 2014), xii.

225 **These laborers had little bargaining power**: Wise and Trist, "Buyer Power," 4–8; Lance Compa, *Blood, Sweat, and Fear: Workers' Rights in U.S. Meat and Poultry Plants* (New York: Human Rights Watch, 2004).

225 **It was no accident that the hog industry:** Deborah Fink, *Cutting into the Meat-packing Line* (Chapel Hill: University of North Carolina Press, 1998), 1–2, 50–69; Brian Page, "Restructuring Pork Production, Remaking Rural Iowa," in *Globalising Food,* ed. David Goodman and Michael Watts (London: Routledge, 1997), 133–157.

225 **For every dollar spent on pork:** Wise and Trist, "Buyer Power," 19.

225 **In constant dollars, the price of pork:** John McGlone, "Swine," in *Animal Welfare in Animal Agriculture,* ed. Wilson G. Pond, Fuller W. Bazer, and Bernard E. Rollin (Boca Raton: CRC Press, 2012), 149.

226 **"It came through the woods":** "Huge Spill of Hog Waste Fuels an Old Debate in North Carolina," *New York Times,* June 25, 1995.

226 **A 250-pound hog excretes 7.8 pounds of feces:** Al Jensen, *Management and Housing for Confinement Swine Production* (Urbana: University of Illinois, 1972), 19; Pew Commission, *Putting Meat,* 29.

226 **Strict rules governed the disposal of human waste:** *Quarterly Hogs and Pigs* (Washington, DC: US Department of Agriculture, September 29, 1995).

226 **As one man who lived near a manure lagoon explained:** Jeff Tietz, "Boss Hog," *Rolling Stone,* December 14, 2006; also see Joby Warrick and Pat Stith, "New Studies Show That Lagoons Are Leaking," *Raleigh News and Observer,* February 19, 1995; Key and McBride, *Changing Economics,* 12; Darrell Smith, "Can Pigs and People Live in Peace?," *Farm Journal* 119 (1995): 18.

226 **In 2011, at a farm in northern Iowa:** Sarah Zhang, "The Curious Case of the Exploding Pig Farms," *Nautilus,* December 2, 2013.

227 **As a result, the industry shifted to the high plains:** Wise and Trist, "Buyer Power," 7.

227 **All told, experts suggested that American taxpayers:** Elanor Starmer and Timothy Wise, *Living High on the Hog* (Medford, MA: Tufts University, 2007); Elanor Starmer and Timothy Wise, "Feeding at the Trough: Industrial Livestock Firms Saved $35 Billion from Low Feed Prices," *GDAE Policy Brief* 07–03 (2007).

227 **Depending on whose estimate you believe:** Robert Goodland and Jeff Anhang, "Livestock and Climate Change," *World Watch* 22 (2009): 10; Pierre Gerber, *Tackling Climate Change Through Livestock* (Rome: Food and Agriculture Organization of the United Nations, 2013), xii.

227 **More than three-quarters of the antibiotics:** "Record-High Antibiotic Sales for Meat and Poultry Production," Pew Charitable Trusts, http://www. pewtrusts.org/en/multimedia/data-visualizations/2013/recordhigh-antibiotic-sales-for-meat-and-poultry-production; Maryn McKenna, "Imagining the Post-antibiotics Future," *Medium,* November 20, 2013, https://medium.com/@fernnews/imagining-the-post-antibiotics-future-892b57499e77 (accessed June 1, 2014).

227 **according to the Food and Drug Administration:** US Food and Drug Administration, "Antimicrobials Sold or Distributed for Use in Food-Producing Animals" (Washington, DC, September 2014), 16–17.

228 **In response to such dangers:** "Pig Out," *Nature* 486 (2012): 440.

228 **Pork producers in the United States:** "Antimicrobials/Antibiotics," National Pork Producers Council, http://www.nppc.org/issues/animal-health-safety/antimicrobials-antibiotics (accessed October 7, 2014).

228 **"Antibiotic use in food animals":** *Antibiotic Resistance Threats in the United States* (Atlanta: Centers for Disease Control and Prevention, 2013), 11, 14.

228 **The FDA in 2013 issued:** Sabrina Tavernise, "Antibiotics in Livestock," *New York Times*, October 2, 2014.

228 **"They love it":** Matthew Scully, *Dominion* (New York: St. Martin's Press, 2002), 258.

229 **Workers who enter the barns:** Abigail Woods, "Rethinking the History of Modern Agriculture: British Pig Production, c. 1910–65," *Twentieth Century British History* 23 (2011): 177.

229 **"Acute and chronic infections of the respiratory tract in pigs":** Mark Ackermann, "Respiratory Tract," in *Biology of the Domestic Pig*, ed. Wilson G. Pond and Harry J. Mersmann (Ithaca, NY: Cornell University Press, 2001), 527.

229 **But in a confinement facility with metal bars and concrete floors:** Ruth Layton, "Animal Needs and Commercial Needs," in *The Future of Animal Farming*, ed. Marian Dawkins and Roland Bonney (Malden, MA: Blackwell, 2008), 88–92; "Early Weaned Behavior May Last Lifetime," *National Hog Farmer*, December 1998.

229 **"Without malleable substrates to chew":** Stanley Curtis, Sandra Edwards, and Harold Gonyou, "Ethology and Psychology," in Pond, *Biology of the Domestic Pig*, 66–67; Harry Blokhuis et al., *Scientific Report on the Risks Associated with Tail Biting in Pigs* (Parma, Italy: European Food Safety Authority, 2007).

229 **"Once a building is built":** John McGlone, "Alternative Sow Housing Systems" (paper presented at the annual meeting for the Manitoba Pork Producers, Winnipeg, Manitoba, Canada, January 2001), 4.

230 **From standing so long on hard floors:** Joe Vansickle, "Sow Lameness Underrated," *National Hog Farmer*, June 15, 2008.

230 **They cannot groom themselves or interact:** Joe Vansickle, "Sow Housing Debated," *National Hog Farmer*, August 15, 2007; "Crateless Farrowing," *Pig Farming*, January 1997.

230 **With no outlets for natural instincts:** Curtis, Edwards, and Gonyou, "Ethology and Psychology," 67. Also see Donald Broom and Andrew Fraser, *Domestic Animal Behaviour and Welfare* (Wallingford, UK: CABI, 2007), 275.

230 **But they keep trying:** Layton, "Animal Needs," 90.

231 **"Contemporary swine production systems may create frustration":** Curtis, Edwards, and Gonyou, "Ethology and Psychology," 52.

CHAPTER 18

233 **Cattle ranchers, he had found, "cared deeply":** Bernard Rollin, *Putting the Horse Before Descartes* (Philadelphia: Temple University Press, 2011), 213.

234 **In his talk to the farmers:** Rollin, *Putting the Horse*, 213.

234 **"I have been feeling lousy":** Rollin, *Putting the Horse*, 213.

234 **On the soundtrack, Willie Nelson sings a Coldplay song:** Johnny Kelly, dir., *Back to the Start* (Chipotle, 2011).

235 **China and Brazil accounted for nearly all of that growth:** *World Agriculture Towards 2015/2030* (Rome: Food and Agriculture Organization of the United Nations, 2002), 58–59.

235 **During that same period, meat production in the developing world increased:** Henning Steinfeld, *Livestock's Long Shadow* (Rome: Food and Agriculture Organization of the United Nations, 2006), 16–17; James Galloway et al., "International Trade in Meat: The Tip of the Pork Chop," *Ambio* 36 (2007): 622–629.

235 **Thanks to genetically modified seeds, fertilizers, pesticides, and herbicides:** Steinfeld, *Livestock's Long Shadow*, 12.

236 **It functions much like the federal oil reserve:** Mindi Schneider, *Feeding China's Pigs* (Minneapolis, MN: Institute for Agriculture and Trade Policy, 2011), 3.

236 **Though China is still largely self-sufficient in pork:** David Bracken, "Chinese Company to Acquire Smithfield Foods for $4.7 Billion," *Raleigh News and Observer*, May 29, 2013.

236 **Those foreign soybeans do not get turned into tofu:** Leslie Hook and Emiko Terazono, "China's Appetite for Food Imports to Fuel Agribusiness," *Financial Times*, June 6, 2013.

236 **Two decades later, that figure had dropped:** Schneider, *Feeding China's Pigs*, 3–6.

239 **After their pig-park study, Stolba and Wood-Gush concluded:** A. Stolba and D. G. M. Wood-Gush, "The Behaviour of Pigs in a Semi-natural Environment," *Animal Production* 48 (1989): 423. Also see Alex Stolba and D. G. Wood-Gush, "The Identification of Behavioural Key Features and Their Incorporation into a Housing Design for Pigs," *Annals of Veterinary Research* 15 (1983): 297.

239 **If breeders selected for maternal abilities as well as rapid weight gain:** A. Kittawornrat and J. J. Zimmerman, "Toward a Better Understanding of Pig Behavior and Pig Welfare," *Animal Health Research Reviews* 12 (2011): 25–32.

240 **In response to *Animal Machines*, the British government formed the Brambell Commission:** H. van de Weerd and V. Sandilands, "Bringing the Issue of Animal Welfare to the Public: A Biography of Ruth Harrison (1920–2000)," *Applied Animal Behaviour Science* 113 (2008): 404–410.

240 **The inquiry uncovered appalling conditions:** Ruth Harrison et al., *Animal Machines* (Boston: CABI, 2013), 11.

240 **A group called the Farm Animal Welfare Council later revised the five freedoms:** Harrison et al., *Animal Machines*, 12.

240 **These recommendations carried no legal weight:** I. Veissier et al., "European Approaches to Ensure Good Animal Welfare," *Applied Animal Behaviour Science* 113 (2008): 279–297.

240 **In 1997 an EU veterinary committee issued a 190-page report:** European Union, *The Welfare of Intensively Kept Pigs* (Brussels: European Commission, 1997), 8.73.

240 **The farthest-reaching provision banned the use of gestation crates:** "Animal Welfare on the Farm: Pigs," Council Directive 2001/88/EC, October 23, 2001.

240 **Canada, too, has since ordered that gestation crates:** *Code of Practice for the Care and Handling of Pigs* (Ottawa, ON: National Farm Animal Care Council, 2014), 10–12.

241 **Promoting farm animal welfare has proved more difficult:** J. A. Mench, "Farm Animal Welfare in the U.S.A.," *Applied Animal Behaviour Science* 113 (2008): 298–312.

241 **In 2008 the Pew Charitable Trusts, a prominent nonprofit:** Pew Commission on Industrial Farm Animal Production, *Industrial Food Production in America: Examining the Impact of the Pew Commission's Priority Recommendations* (Philadelphia: Pew Charitable Trusts, 2013), 26.

241 **"Gestation crates are a real problem"**: Humane Society of the United States (HSUS), *Undercover at Smithfield Foods* (Washington, DC: HSUS, 2010), 2.

241 **A number of US states have banned gestation crates:** Humane Society of the United States (HSUS), *Welfare Issues with Gestation Crates for Pregnant Sows* (Washington, DC: HSUS, 2013), 1–2.

241 **Early in 2014, Smithfield promised:** Christopher Doering, "Smithfield Urges Farmers to End Use of Gestation Crates," *USA Today*, January 7, 2014; Mike Hughlett, "Consumer Pressure Leads Cargill to Give Pigs More Room," *Minneapolis Star Tribune*, June 8, 2014.

242 **"Their feelings aren't rational":** Lydia Depillis, "Big Agriculture Wants to Reach Millennials," *Washington Post*, May 14, 2014.

242 **In the words of one European official:** Richard Perren, *Taste, Trade and Technology* (Burlington, VT: Ashgate, 2006), 203; also see Joe Vansickle, "Pork Board Reacts to Welfare Demands," *National Hog Farmer*, February 15, 2002.

242 **The EU, for instance, has a stricter standard for "organic" pork:** "Organically Grown Agricultural Products and Foodstuffs," European Commission, October 6, 2008, http://europa.eu/legislation_summaries/other/l21118_en.htm (accessed February 23, 2014).

242 **Producers in Denmark have created a special category called the "welfare pig":** "New Welfare Pig for Danish Market," *The Meat Site*, October 25, 2013, http://www.themeatsite.com/meatnews/22915/new-welfare-pig-for-danish-market (accessed October 29, 2014).

242 **Similar standards in the United Kingdom qualify a pig as "Freedom Food":** "Major Milestone for Welfare as McDonald's Announce Switch to 100% Freedom Food Pork," Freedom Food, April 2013, http://www.freedomfood.co.uk/news/2013/04/mcdonalds (accessed November 13, 2013).

242 **The American Humane Association adopted the British Freedom Food standards:** "Du Breton Natural Pork Earns Animal Welfare Certification," *National Hog Farmer*, May 15, 2001.

242 **The meat case at every Whole Foods store:** "An Inside Look," Global Animal Partnership, http://www.globalanimalpartnership.org/about-us/an-inside-look (accessed February 21, 2014).

243 **The US Department of Agriculture maintains its own standards:** "Meat and Poultry Labeling Terms," US Department of Agriculture, October 24, 2014, http://www.fsis.usda.gov/wps/portal/fsis/topics/food-safety-education/get-answers/food-safety-fact-sheets/food-labeling/meat-and-poultry-labeling-terms/meat-and-poultry-labeling-terms (accessed February 23, 2014); Katie Abrams, Courtney Meyers, and Tracy Irani, "Naturally Confused: Consumers' Perceptions of All-Natural and Organic Pork Products," *Agriculture and Human Values* 27 (2010): 365–374.

243 **These certifications serve as marketing tools:** P. C. Thompson et al., "Livestock Welfare Product Claims," *Journal of Animal Science* 85 (2007): 2354–2360.

243 **Confinement farming is "not better for the animals":** Jonathan Safran Foer, *Eating Animals* (New York: Little, Brown, 2009), 167.

244 **Eventually, his network grew to hundreds of farms:** Nicolette Hahn Niman, *Righteous Porkchop* (New York: Collins Living, 2009), 116–125; M. S. Honeyman et al., "The United States Pork Niche Market Phenomenon," *Journal of Animal Science* 84 (2006): 2269–2275.

244 **"All our hogs are raised outdoors"**: "All-Natural Pork," Niman Ranch, http://www.nimanranch.com/pork.aspx (accessed November 21, 2013).

244 **EcoFriendly Foods, for example, advertises that all of its pigs**: "Pigs," EcoFriendly Foods, http://www.ecofriendly.com/pigs (accessed November 17, 2013).

245 **A study at Iowa State University estimated**: Honeyman et al., "United States Pork Niche."

245 **More recent statistics are not available**: "Hog Protocols," Niman Ranch, http://www.nimanranch.com/Protocols.aspx (accessed November 17, 2013).

246 **"Good welfare means that the base price of pork will inevitably rise"**: Ruth Layton, "Animal Needs and Commercial Needs," in *The Future of Animal Farming*, ed. Marian Dawkins and Roland Bonney (Malden, MA: Blackwell, 2008), 90.

246 **In fully pastured systems, with slower-growing heritage breeds**: Honeyman et al., "United States Pork Niche," 2272.

246 **Then his partners process and cure the meat**: Julie Robinson, "Porcine Perfection," *Charleston Gazette*, October 1, 2011; C. W. Talbott et al., "Enhancing Pork Flavor and Fat Quality with Swine Raised in Sylvan Systems," *Renewable Agriculture and Food Systems* 21 (2006): 183–191; Chuck Talbott et al., "Potential for Small-Scale Farmers to Produce Niche Market Pork Using Alternative Diets, Breeds and Rearing Environments," *Renewable Agriculture and Food Systems* 19 (2007): 135–140; Peter Kaminsky, *Pig Perfect* (New York: Hyperion, 2005).

246 **Talbott explains the dilemma of modern pork**: Chuck Talbott, interview with author, October 15, 2012.

EPILOGUE

247 **When presented with a salad or burrito**: Technomic, *Center of the Plate: Beef and Pork Consumer Trend Report* (Chicago: Technomic, 2013).

247 **Recent figures from the US Department of Agriculture show**: Christopher Davis and Biing-Hwan Lin, *Factors Affecting U.S. Pork Consumption* (Washington, DC: US Department of Agriculture, 2005).

248 **In 2013 the activist group Mercy for Animals released footage**: Mercy for Animals, "Walmart Pork Supplier Caught Abusing Mother Pigs and Piglets," YouTube, October 29, 2013, http://www.youtube.com/watch?feature=player_embedded&v=-KoVAkgPexU (accessed February 24, 2014).

248 **Recently, however, as cognitive and behavioral scientists have confirmed**: Donald M. Broom, "Cognitive Ability and Awareness in Domestic Animals and Decisions About Obligations to Animals," *Applied Animal Behaviour Science* 126 (2010): 1–11.

248 **British chef Hugh Fearnley-Whittingstall calls pigs**: Hugh Fearnley-Whittingstall, *The River Cottage Meat Book* (Berkeley, CA: Ten Speed Press, 2007), 112.

249 **Activist Gail Eisnitz, who has investigated all types of animal cruelty**: Nicolette Hahn Niman, *Righteous Porkchop* (New York: Collins Living, 2009), 215.

249 **These sows, Jonathan Safran Foer writes in *Eating Animals***: Jonathan Safran Foer, *Eating Animals* (New York: Little, Brown, 2009), 183.

249 **That means that sows, after giving birth**: Matthew Scully, *Dominion* (New York: St. Martin's Press, 2002), 266.

249 **"The pigs are treated like shit":** Sarah Hepola, "A Wonderful, Magical Animal," *Salon*, July 11, 2008, http://www.salon.com/2008/07/11/magical_animal (accessed October 29, 2014).

249 **A few animal scientists have proposed:** Adam Shriver, "Not Grass-Fed, but at Least Pain-Free," *New York Times*, February 18, 2010.

250 **That's the position taken by Cromwell:** Mercy for Animals, "Walmart Pork Supplier."

250 **In addition to working with the more familiar heritage breeds:** Peter Kaminsky, *Pig Perfect* (New York: Hyperion, 2005).

250 **The Roman historian Livy noted in the second century BC:** Emily Gowers, *The Loaded Table* (New York: Oxford University Press, 1993), 51.

251 **No restaurants have resurrected the Roman recipe for roast udder:** Dana Goodyear, *Anything That Moves* (New York: Riverhead, 2013), 74.

251 **It sold well:** Stephanie Strom, "Demand Grows for Hogs That Are Raised Humanely Outdoors," *New York Times*, January 20, 2014.

251 **Even as pork marketers flogged lean pork chops:** David Sax, "The Bacon Boom Was Not an Accident," *Businessweek*, October 6, 2013.

251 **sales of lean pork stayed flat:** J. L. Anderson, "Lard to Lean: Making the Meat-Type Hog in Post–World War II America," in *Food Chains*, ed. Warren Belasco and Roger Horowitz (Philadelphia: University of Pennsylvania Press, 2009).

251 **A couple of cooks in Kansas City found Internet fame:** Jason, "Bacon Explosion: The BBQ Sausage Recipe of all Recipes," BBQ Addicts, December 23, 2008, http://www.bbqaddicts.com/blog/recipes/bacon-explosion (accessed February 24, 2014).

252 **Dozens of cities hosted bacon festivals:** Erin Zimmer, "Bacon Bra," Serious Eats, April 2, 2008, http://www.seriouseats.com/2008/04/bacon-bra-brassiere-womens -edible-underwear.html (accessed February 24, 2014).

252 **It's an indulgence, they write:** Joshua Applestone, Jessica Applestone, and Alexandra Zissu, *The Butcher's Guide to Well-Raised Meat* (New York: Clarkson Potter, 2011), 130.

253 **"We have been raising happy, healthy pigs since 1994":** "Why Pastured Pork," Wil-Den Family Farms, http://www.wildenfamilyfarms.com/Main/pasturedpork .html (accessed November 17, 2013).

253 **EcoFriendly Farms reduces it to an equation:** EcoFriendly Farms, "Pigs." Also see C. W. Talbott et al., "Enhancing Pork Flavor and Fat Quality with Swine Raised in Sylvan Systems," *Renewable Agriculture and Food Systems* 21 (2006): 183–191.

253 **If public concern drives further agricultural reforms:** N. Pelletier et al., "Life Cycle Assessment of High-and Low-Profitability Commodity and Deep-Bedded Niche Swine Production Systems in the Upper Midwestern United States," *Agricultural Systems* 103 (2010): 599–608.

255 **"You may end up paying twice as much":** Fearnley-Whittingstall, *River Cottage Meat Book*, 108.

255 **The farmers' markets and upscale grocers:** Brad Weiss, "Configuring the Authentic Value of Real Food," *American Ethnologist* 39 (2012): 615–616, 623–624.

256 **The majority of people, the study concluded:** Jayson L. Lusk, F. Bailey Norwood, and Robert W. Prickett, "Consumer Preferences for Farm Animal Welfare" (working paper, Department of Agricultural Economics, Oklahoma State University, 2007), 13.

257 **In historical terms, Americans now spend a tiny portion:** Cynthia Northrup, *The American Economy* (Santa Barbara, CA: ABC-CLIO, 2011), 102.

257 **According to John McGlone, an agriculture professor:** John McGlone, "Swine," in *Animal Welfare in Animal Agriculture*, ed. Wilson G. Pond, Fuller W. Bazer, and Bernard E. Rollin (Boca Raton: CRC Press, 2012), 149.

INDEX